Non-Newtonian Calculus

Michael Grossman
(Lowell Technological Institute)

Robert Katz

1972

Lee Press

Pigeon Cove, Massachusetts 01966

Non-Newtonian calculus [by] Michael Grossman [and] Robert Katz. Pigeon Cove, Mass., Lee Press, 1972.

viii, 94 p. 23 cm.

1. Calculus.   I. Katz, Robert, 1922–   joint author.
II. Title.

QA303.G88   515   72–76466
ISBN 0–912938–01–3   MARC

Library of Congress   72 [4]

Fourth Printing, July 1978

Manufactured in the United States of America

<u>F o r e w o r d</u>

We are pleased to present here the results of our investigations into the non-Newtonian calculi, which are markedly different from the classical calculus developed by Newton and Leibniz three centuries ago.

Robert Katz

Michael Grossman

84 Main Street
Rockport, Massachusetts
December 14, 1971

In the classical calculus the line is used as a standard against which other curves are compared. In the following remarks, Roger Joseph Boscovich (1711 - 1787) may have been the only person to anticipate a fundamental idea in non-Newtonian calculus: *a nonlinear curve may be used as a standard against which other curves (including lines) can be compared.*

'But if some mind very different from ours were to look upon some property of some curved line as we do on the evenness of a straight line, he would not recognize as such the evenness of a straight line; nor would he arrange the elements of his geometry according to that very different system, and would investigate quite other relationships as I have suggested in my notes.

'We fashion our geometry on the properties of a straight line because that seems to us to be the simplest of all. But really all lines that are continuous and of a uniform nature are just as simple as one another. Another kind of mind which might form an equally clear mental perception of some property of any one of these curves, as we do of the congruence of a straight line, might believe these curves to be the simplest of all, and from that property of these curves build up the elements of a very different geometry, referring all other curves to that one, just as we compare them to a straight line. Indeed, these minds, if they noticed and formed an extremely clear perception of some property of, say, the parabola, would not seek, as our geometers do, to *rectify* the parabola, they would endeavour, if one may coin the expression, to *parabolify* the straight line.'

The quotation above appears in Dr. J. F. Scott's article "Boscovich's Mathematics," which is one of nine articles in *Roger Joseph Boscovich,* edited by Lancelot Law Whyte and first published by George Allen & Unwin in 1961.

<u>P R E F A C E</u>

In August of 1970 we constructed a comprehensive family
of calculi, which includes the classical calculus as well as
an infinite subfamily of non-Newtonian calculi that we had
constructed three years earlier.

Each calculus possesses the following:

a distinctive method of measuring changes in
function arguments;

a distinctive method of measuring changes in
function values;

four operators: a gradient (i.e., an average
rate of change), a derivative, a natural
average, and an integral;

a characteristic class of functions having a
constant derivative;

a Basic Theorem involving the gradient, derivative, and natural average;

a Basic Problem whose solution motivates a simple definition of the integral in terms of
the natural average;

and two Fundamental Theorems which reveal that
the derivative and integral are 'inversely'
related in an appropriate sense.

A popular method of creating a new mathematical system
is to vary the axioms of a known system. Although it may be
possible to axiomatize the classical calculus, we did not and
shall not pursue that course. Instead, in Chapter 1 we for-
mulate the classical calculus in a novel manner that leads
naturally to the subsequent construction of the non-Newtonian
calculi.

In Chapters 2 - 4, we construct three specific non-Newtonian calculi: the geometric, anageometric, and bigeometric.
Chapter 5 is devoted to arithmetics (slightly specialized complete ordered fields), which are used in subsequent chapters.
In Chapter 6 we simultaneously construct all the calculi, indicate the uniform relationships between the corresponding
operators of any two calculi, and provide suitable graphic

interpretations. Chapters 7 and 8 contain brief developments
of the quadratic, anaquadratic, biquadratic, harmonic, anaharmonic, and biharmonic calculi. (The harmonic calculus is not
to be confused with harmonic analysis.) Chapter 9 includes
a variety of heuristic guides for selecting gradients, derivatives, averages, and integrals. In Chapter 10 we discuss
certain generalized spaces, vectors, and least squares methods,
which were suggested by our work with the non-Newtonian calculi; we introduce a trend concept that may be considered as a
kind of global derivative; and we indicate some connections
between the non-Newtonian calculi and calculus in Banach
spaces. Sundry digressions and comments have been placed in
the Notes at the rear of the book; for example, in Note 2 we
explain how a simple algebraic identity suggested the possibility of constructing the geometric calculus.

Since this book is intended for a wide audience, including students, engineers, scientists, as well as mathematicians,
we have presented many details that would not appear in a re-
search report and we have excluded all proofs. (All stated
results can be proved in a straightforward way.) It is assumed, of course, that the reader has a working knowledge of
the rudiments of classical calculus.

Finally we wish to express our gratitude to Professor
Dirk J. Struik for his interest and encouragement, and to
Messrs. Charles Rockland, Charles K. Wilkinson, and David J.
Liben for their helpful comments and assistance. However,
we alone are responsible for the form and content of this
book.

Criticism and suggestions are cordially invited.

M.G.

R.K.

C O N T E N T S

PRELIMINARIES

Since complex numbers are not used in this book, the
word "number" will mean real number. Except for f, g, and
h, lower-case Roman letters are used for referring to numbers. The letter R denotes the set of all numbers, and the
symbol R_+ denotes the set of all positive numbers.

If $r < s$, then the _interval_ [r,s] is the set of all
numbers x such that $r \leq x \leq s$. (We do not use open intervals in this book.) The _interior_ of [r,s] is obtained by
deleting r and s therefrom. The _classical extent_ of [r,s]
is $s - r$. (Although the number $s - r$ is usually called the
length of [r,s], we use the term "classical extent" in order
to achieve a uniform terminology for all the calculi.)

By a _point_ we mean an ordered pair of numbers; by a
function we mean a set of points, each distinct two of which
have distinct first members. We use the letters f, g, h,
and occasionally others, for referring to functions, and we
usually identify a function with its graph, thereby avoiding undesirable circumlocution.

The _domain_ of a function is the set of all its _argu-
ments_ (first members of the points); the _range_ of a function
is the set of all its _values_ (second members of the points).
A function is said to be _on_ its domain, to be _onto_ its range,
and to be _defined_ at each of its arguments.

If every two distinct points of a function f have distinct second members, then f is _one-to-one_ and its _inverse_,
f^{-1}, is the one-to-one function consisting of all points
(y,x) for which (x,y) is a point of f. Please bear in mind
that f^{-1} is not the reciprocal of f.

A _positive function_ is a function whose values are all
positive; a _discrete function_ is a function that has only a
finite number of arguments.

A _linear function_ is a function of the form $mx + c$,
where m and c are constants and x is unrestricted in R. An
especially important linear function is the _identity func-
tion_, I, for which $I(x) = x$. (All linear functions are on
R.) The function exp assigns to each number x the number e^x,
where e is the base of the natural logarithm function, ln,

which is the inverse of exp.

In the sequel we shall introduce some new terms and symbols, which, however, we have endeavored to keep to a minimum.

Chapter 1

THE CLASSICAL CALCULUS

1.1 INTRODUCTION

In this chapter we present some of the basic ideas of the classical calculus in a novel manner that leads naturally to the subsequent construction of the non-Newtonian calculi.

In the classical calculus, changes in function arguments and values are measured by differences.

1.2 THE CLASSICAL GRADIENT

The classical change of a function f over [r,s] (or from r to s) is the number f(s) - f(r).

A classically-uniform function is a function that is on R, is continuous, and has the same classical change over any two intervals of equal classical extent. Clearly every constant function on R is classically-uniform.

Since the classically-uniform functions turn out to be the linear functions, we shall henceforth use the term "linear function."

It is characteristic of a linear function that for each arithmetic progression of arguments, the corresponding sequence of values is also an arithmetic progression.

The classical slope of a linear function is its classical change over any interval of classical extent 1. Of course, the classical slope of the function mx + c turns out to be m.

The classical gradient of a function f over [r,s] is the classical slope of the linear function containing the points (r,f(r)) and (s,f(s)), and turns out to be

$$\frac{f(s) - f(r)}{s - r} .$$

There are several reasons why we decided to use the British term "gradient" instead of "average rate of change" or "rate of change": the first term is the shortest; the second leaves us wondering of what it is the average; the third is often taken to mean "instantaneous rate of change."

There will be no confusion with the gradient concept in vector analysis, since vector analysis is not treated in this book.

Clearly the classical gradient is independent of the origins used for measuring the magnitudes that the function arguments and values may represent.

1.3 THE CLASSICAL DERIVATIVE

Throughout this section, f is assumed to be a function defined at least on an interval containing the number a in its interior.

If the following limit exists, we denote it by [Df](a), call it the <u>classical</u> <u>derivative</u> of f at a, and say that f is classically differentiable at a:

$$\lim_{x \to a} \frac{f(x) - f(a)}{x - a} \ .$$

The classical derivative of f, denoted by Df, is the function that assigns to each number t the number [Df](t), if it exists.

The operator D is additive and homogeneous:

$$D(f + g) = Df + Dg,$$

$$D(c \cdot f) = c \cdot Df, \quad c \text{ constant.}$$

(We are intentionally avoiding the term "linear operator.")

In the classical calculus a distinctive role is played by the linear functions, for only they possess classical derivatives that are constant on R.

Next we define a familiar concept in a special way that will permit a suitable generalization in Section 6.3.

The <u>classical</u> <u>tangent</u> to a function f at the point (a,f(a)) is the unique linear function g, if it exists, which possesses the following two properties.

1. The linear function g contains (a,f(a)).
2. For each linear function h containing (a,f(a))
   and distinct from g, there is a positive number
   p such that for every number x in [a - p, a + p]
   but distinct from a,

$$\left| g(x) - f(x) \right| < \left| h(x) - f(x) \right| .$$

Mr. Charles Rockland of Princeton University has proved that $[Df](a)$ exists if and only if f has a classical tangent at $(a, f(a))$, and that if $[Df](a)$ does exist, it equals the classical slope of that classical tangent.

We say that two functions are _classically tangent_ at a common point if and only if they have the same classical tangent there.

1.4 THE ARITHMETIC AVERAGE

The _arithmetic average_ of n numbers $v_1, \ldots, v_n$ is the number $(v_1 + \ldots + v_n)/n$.

In our treatment of classical calculus an important role is played by the arithmetic average of a continuous function on an interval, the definition of which will be simplified by introducing the concept of an arithmetic partition of an interval.

An _arithmetic partition_ of an interval $[r, s]$ is any arithmetic progression whose first and last terms are r and s, respectively. An arithmetic partition is _n-fold_ if it has n terms.

The _arithmetic average_ of a continuous function f on $[r, s]$ is denoted by

$$M_r^s f$$

and defined to be the limit of the convergent sequence whose nth term is the arithmetic average of $f(a_1), \ldots, f(a_n)$, where $a_1, \ldots, a_n$ is the n-fold arithmetic partition of $[r, s]$.

The operator M_r^s is additive and homogeneous:

$$M_r^s (f + g) = M_r^s f + M_r^s g,$$

$$M_r^s (c \cdot f) = c \cdot M_r^s f, \quad c \text{ constant.}$$

The arithmetic average of a linear function on $[r, s]$ is equal to the arithmetic average of its values at r and s, and is equal to its value at the arithmetic average of r and s.

The arithmetic average is the only operator that has the following three properties.

1. For any interval $[r,s]$ and any constant function $h(x) = b$ on $[r,s]$,

$$M_r^s h = b.$$

2. For any interval $[r,s]$ and any functions f and g that are continuous on $[r,s]$, if $f(x) \leq g(x)$ for every number x in $[r,s]$, then

$$M_r^s f \leq M_r^s g.$$

3. For any numbers r,s,t such that $r < s < t$, and any function f continuous on $[r,t]$,

$$(s - r) \cdot M_r^s f + (t - s) \cdot M_s^t f = (t - r) \cdot M_r^t f.$$

1.5 THE BASIC THEOREM OF CLASSICAL CALCULUS

Our discussion of the Basic Theorem of Classical Calculus begins with its discrete analogue, which is a proposition that concerns discrete functions and appropriately conveys the spirit of the theorem.

The Discrete Analogue of the Basic Theorem of Classical Calculus

Let h be a discrete function whose arguments $a_1, \ldots, a_n$ are an arithmetic partition of $[r,s]$. Then the arithmetic average of the following $n - 1$ classical gradients is equal to the classical gradient of h over $[r,s]$:

$$\frac{h(a_{i+1}) - h(a_i)}{a_{i+1} - a_i}, \quad i = 1, \ldots, n-1.$$

The Basic Theorem of Classical Calculus

If Dh is continuous on $[r,s]$, then its arithmetic average on $[r,s]$ equals the classical gradient of h over $[r,s]$, that is

$$M_r^s(Dh) = \frac{h(s) - h(r)}{s - r}.$$

In view of the preceding theorem we say that the arithmetic average fits naturally into the scheme of classical calculus.

The Basic Theorem of Classical Calculus provides an immediate solution to the following problem.

The Basic Problem of Classical Calculus

Suppose that the value of a function h is known at an argument r, and suppose that f, the classical derivative of h, is continuous and known at each number in [r,s]. Find h(s).

Solution

By the Basic Theorem of Classical Calculus we have

$$M_r^s f = M_r^s(Dh) = \frac{h(s) - h(r)}{s - r}.$$

Solving for h(s), we get

$$h(s) = h(r) + (s - r) \cdot M_r^s f.$$

The number $(s - r) \cdot M_r^s f$ in the foregoing solution arises with sufficient frequency to warrant a special name, which is introduced in the next section and is justified by the ensuing discussion.

1.6 THE CLASSICAL INTEGRAL

The classical integral of a continuous function f on [r,s] is the number $(s - r) \cdot M_r^s f$, and is denoted by $\int_r^s f$. We set $\int_r^r f = 0$.

The classical integral is a weighted arithmetic average, since

$$\int_r^s f = M_r^s[(s - r) \cdot f].$$

Furthermore, $\int_r^s f$ equals the limit of the convergent sequence whose nth term is the sum $k_n \cdot f(a_1) + \ldots + k_n \cdot f(a_{n-1})$, where $a_1, \ldots, a_n$ is the n-fold arithmetic partition of [r,s], and k_n is the common value of

$$a_2 - a_1, \; a_3 - a_2, \; \ldots, \; a_n - a_{n-1}.$$

The operator $\displaystyle\int_r^s$ is additive and homogeneous:

$$\int_r^s (f + g) = \int_r^s f + \int_r^s g,$$

$$\int_r^s (c \cdot f) = c \cdot \int_r^s f, \quad c \text{ constant.}$$

The classical integral is the only operator that has the following three properties.

1. For any interval $[r,s]$ and any constant function $h(x) = b$ on $[r,s]$,

$$\int_r^s h = (s - r) \cdot b.$$

2. For any interval $[r,s]$ and any functions f and g that are continuous on $[r,s]$, if $f(x) \leq g(x)$ for every number x in $[r,s]$, then

$$\int_r^s f \leq \int_r^s g.$$

3. For any numbers r, s, t such that $r < s < t$, and any function f continuous on $[r,t]$,

$$\int_r^s f + \int_s^t f = \int_r^t f.$$

1.7 THE FUNDAMENTAL THEOREMS OF CLASSICAL CALCULUS

The classical derivative and integral are 'inversely' related in the sense indicated by the following two theorems.

First Fundamental Theorem

If f is continuous on $[r,s]$, and

$$g(x) = \int_r^x f, \quad \text{for every number } x \text{ in } [r,s],$$

then

$$Dg = f, \quad \text{on } [r,s].$$

Second Fundamental Theorem

If Dh is continuous on $[r,s]$, then

$$\int_r^s (Dh) = h(s) - h(r).$$

Chapter 2
THE GEOMETRIC CALCULUS

2.1 INTRODUCTION

During the Renaissance many scholars, including Galileo, discussed the following problem:

> Two estimates, 10 and 1000, are proposed as
> the value of a horse. Which estimate, if any,
> deviates more from the true value of 100?

The scholars who maintained that the deviations should be measured by differences concluded that the estimate of 10 was closer to the true value. However, Galileo eventually maintained that the deviations should be measured by ratios, and he concluded that the two estimates deviated equally from the true value.

Now let us consider a more sophisticated problem. At time r, a man invests f(r) dollars with a promoter who guarantees that at a certain subsequent time s, the value of the investment would be f(s) dollars. In event that the investor should desire to withdraw at any other time t, it was agreed that the value of the investment increases continuously and uniformly. The problem is this: How much, f(t), would the investor be entitled to at time t? We shall give two reasonable solutions; there is no unique solution.

Solution 1. Since it was agreed that the value of the investment increases uniformly, we may reasonably assume that it increases by equal amounts in equal times. Since it was agreed, furthermore, that the value of the investment increases continuously, it can be proved that the growth must be linear and, in fact, that

$$f(t) = f(r) + \left[\frac{f(s) - f(r)}{s - r} \right] (t - r).$$

Notice that the expression within the brackets represents the classical gradient of f over [r,s].

Solution 2. Since it was agreed that the value of the investment increases uniformly, we may reasonably assume that it in-

9

creases by equal percents in equal times. Since, furthermore,
the value of the investment increases continuously, it can be
proved that the growth must be exponential and, in fact, that

$$f(t) = f(r) \cdot \left[\left(\frac{f(s)}{f(r)} \right)^{\frac{1}{s-r}} \right]^{t-r} .$$

We shall see that the expression within the brackets is of
fundamental importance in geometric calculus.

In the geometric calculus, changes in function arguments
and values are measured by differences and ratios, respectively, and the operators are applied only to positive functions. Therefore, every function in this chapter is positive,
except for the natural logarithm function, ln, which is used
for special purposes. Lest the reader be disappointed by
such a restriction, we hasten to say that a geometric-type
calculus for negative (valued) functions is discussed briefly
in Section 6.10.

Heuristic guides for selecting appropriate gradients,
derivatives, averages, and integrals are discussed in Chapter
9.

2.2 THE GEOMETRIC GRADIENT

The geometric change of a positive function f over [r,s]
is the number f(s)/f(r).

A geometrically-uniform function is a positive function
that is on R, is continuous, and has the same geometric change
over any two intervals of equal classical extent. Clearly
every positive constant function on R is geometrically-uniform.

The geometrically-uniform functions are those expressible
in the form exp(mx + c), where m and c are constants and x is
unrestricted in R. If p is a positive constant, then the functions p^{mx+c} are also geometrically-uniform, for they are expressible in the form indicated.

It is characteristic of a geometrically-uniform function
that for each arithmetic progression of arguments, the corresponding sequence of values is a geometric progression. That
fact provides one reason for the name "geometric calculus";
another reason will appear in Section 2.4.

The geometric slope of a geometrically-uniform function is its geometric change over any interval of classical extent 1. For example, the geometric slopes of the functions $\exp(mx + c)$ and p^{mx+c} turn out to be e^m and p^m, respectively.

If one plots a geometrically-uniform function h on semi-log paper that is logarithmically scaled on the y-axis, the result is a straight line whose classical slope equals the natural logarithm of the geometric slope of h. Other aspects of graphical interpretation will be discussed more fully in Section 6.11.

The geometric gradient of a positive function f over $[r,s]$ is the geometric slope of the geometrically-uniform function containing $(r,f(r))$ and $(s,f(s))$, and turns out to be

$$\left[\frac{f(s)}{f(r)}\right]^{\frac{1}{s-r}}.$$

Clearly the geometric gradient is independent of the origin and unit used for measuring the magnitudes that the arguments and values may represent, respectively.

The foregoing expression for the geometric gradient yields the indeterminate form 1^{∞} when $r = s$, in contrast to the indeterminate form $0/0$ yielded by the expression for the classical gradient.

2.3 THE GEOMETRIC DERIVATIVE

Throughout this section, f is assumed to be a positive function defined at least on an interval containing the number a in its interior.

If the following limit exists and is positive, we denote it by $[\widetilde{D}f](a)$, call it the geometric derivative of f at a, and say that f is geometrically differentiable at a:

$$\lim_{x \to a} \left[\frac{f(x)}{f(a)}\right]^{\frac{1}{x-a}}.$$

It is possible for the preceding limit to be 0 (this happens, for example, if $f(x) = \exp(-x^{1/3})$ and $a = 0$). Our reason for requiring $[\tilde{D}f](a)$ to be positive will be given in Section 6.10.

It can be proved that $[Df](a)$ and $[\tilde{D}f](a)$ coexist; that is, if either exists then so does the other. Moreover, if they do exist, then $[\tilde{D}f](a)$ equals $\exp\{[Df](a)/f(a)\}$ and equals the geometric slope of the unique geometrically-uniform function that is classically tangent to f at $(a,f(a))$. (Incidentally, the expression $[Df](a)/f(a)$ represents the so-called logarithmic derivative of f at a, which is not a non-Newtonian derivative according to our use of the term.)

The geometric derivative of f, denoted by $\tilde{D}f$, is the function that assigns to each number t the number $[\tilde{D}f](t)$, if it exists.

If f is geometrically-uniform, then $\tilde{D}f$ has a constant value equal to the geometric slope of f. Indeed, only geometrically-uniform functions have geometric derivatives that are constant on R. In particular, if f is a positive constant function on R, then $\tilde{D}f$ is everywhere equal to 1.

Since $\tilde{D}(1 + 1) \neq \tilde{D}(1) + \tilde{D}(1)$, the operator $\tilde{D}$ is not additive; however $\tilde{D}$ is multiplicative:

$$\tilde{D}(f \cdot g) = \tilde{D}f \cdot \tilde{D}g.$$

Furthermore,

$$\tilde{D}(f^c) = (\tilde{D}f)^c, \quad c \text{ constant.}$$

Recalling that the function exp equals its classical derivative, we point out that the function $\exp[\exp(x)]$ equals its geometric derivative.

In Note 1 we define the relative gradient, which is intimately related to the geometric gradient, is often called the compound growth rate, and gives rise to the relative derivative.

2.4 THE GEOMETRIC AVERAGE

The geometric calculus was so named principally because the geometric average fits naturally thereinto, a fact that will be explained in the next section.

The geometric average of n positive numbers $v_1, \ldots, v_n$ is the positive number

$$(v_1 v_2 \cdots v_n)^{1/n}.$$

The <u>geometric</u> <u>average</u> of a continuous positive function f on [r,s] is denoted by $\widetilde{M}_r^s f$ and defined to be the positive limit of the convergent sequence whose nth term is the geometric average of $f(a_1),\ldots,f(a_n)$, where $a_1,\ldots,a_n$ is the n-fold arithmetic partition of [r,s].

The operator $\widetilde{M}_r^s$ is multiplicative:

$$\widetilde{M}_r^s(f \cdot g) = \widetilde{M}_r^s f \cdot \widetilde{M}_r^s g.$$

Also,

$$\widetilde{M}_r^s(f^c) = (\widetilde{M}_r^s f)^c, \quad c \text{ constant.}$$

The geometric average of a geometrically-uniform function on [r,s] is equal to the geometric average of its values at r and s, and is equal to its value at the arithmetic average of r and s.

The geometric average is the only operator that has the following three properties.

1. For any interval [r,s] and any positive constant function h(x) = b on [r,s],

$$\widetilde{M}_r^s h = b.$$

2. For any interval [r,s] and any positive functions f and g that are continuous on [r,s], if $f(x) \le g(x)$ for every number x in [r,s], then

$$\widetilde{M}_r^s f \le \widetilde{M}_r^s g.$$

3. For any numbers r, s, t such that r < s < t, and any positive function f continuous on [r,t],

$$[\widetilde{M}_r^s f]^{s-r} \cdot [\widetilde{M}_s^t f]^{t-s} = [\widetilde{M}_r^t f]^{t-r}.$$

2.5 THE BASIC THEOREM OF GEOMETRIC CALCULUS

Our investigation of non-Newtonian calculi began with an observation of the simple identity indicated in the following proposition. (In Note 2 we explain how this identity provided the original insight into the constructibility of the geometric calculus.)

The Discrete Analogue of the Basic Theorem of Geometric Calculus

Let h be a discrete positive function whose arguments

$a_1, \ldots, a_n$ are an arithmetic partition of $[r,s]$.
Then the geometric average of the following n-1
geometric gradients is equal to the geometric
gradient of h over $[r,s]$:

$$\left[\frac{h(a_{i+1})}{h(a_i)} \right]^{\frac{1}{a_{i+1} - a_i}} , \quad i = 1, \ldots, n-1.$$

The Basic Theorem of Geometric Calculus

If $\widetilde{D}h$ is continuous on $[r,s]$, then its geometric
average on $[r,s]$ equals the geometric gradient
of h over $[r,s]$, that is

$$\widetilde{M}_r^s(\widetilde{D}h) = \left[\frac{h(s)}{h(r)} \right]^{\frac{1}{s-r}} .$$

In view of the preceding theorem we say that the geometric average fits naturally into the scheme of geometric
calculus.

The Basic Theorem of Geometric Calculus provides an
immediate solution to the following problem.

The Basic Problem of Geometric Calculus

Suppose that the value of a positive function h
is known at an argument r, and suppose that f,
the geometric derivative of h, is continuous and
known at each number in $[r,s]$. Find h(s).

Solution

$$h(s) = h(r) \cdot [\widetilde{M}_r^s f]^{s-r} .$$

The number $[\widetilde{M}_r^s f]^{s-r}$ in the foregoing solution arises
with sufficient frequency to warrant a special name, which
is introduced in the next section.

2.6 THE GEOMETRIC INTEGRAL

The geometric integral of a continuous positive func-
tion f on $[r,s]$ is the positive number $[\widetilde{M}_r^s f]^{s-r}$ and is de-

noted by $\widetilde{\int}_r^s f$. We set $\widetilde{\int}_r^r f = 1$.

The geometric integral is a weighted geometric average, since

$$\widetilde{\int}_r^s f = \widetilde{M}_r^s (f^{s-r}).$$

Furthermore, $\widetilde{\int}_r^s f$ equals the positive limit of the convergent sequence whose nth term is the product

$$[f(a_1)]^{k_n} \cdot [f(a_2)]^{k_n} \cdots [f(a_{n-1})]^{k_n},$$

where $a_1, a_2, \ldots, a_n$ is the n-fold arithmetic partition of $[r,s]$, and k_n is the common value of $a_2 - a_1$, $a_3 - a_2, \ldots,$ $a_n - a_{n-1}$.

The operator $\widetilde{\int}_r^s$ is multiplicative:

$$\widetilde{\int}_r^s (f \cdot g) = \widetilde{\int}_r^s f \cdot \widetilde{\int}_r^s g.$$

Also,

$$\widetilde{\int}_r^s (f^c) = \left(\widetilde{\int}_r^s f \right)^c, \quad c \text{ constant.}$$

The geometric integral is the only operator that has the following three properties.

1. For any interval $[r,s]$ and any positive constant function $h(x) = b$ on $[r,s]$,

$$\widetilde{\int}_r^s h = b^{s-r}.$$

2. For any interval $[r,s]$ and any positive functions f and g that are continuous on $[r,s]$, if $f(x) \leq g(x)$ for every number x in $[r,s]$, then

$$\widetilde{\int}_r^s f \leq \widetilde{\int}_r^s g.$$

3. For any numbers r,s,t such that $r < s < t$, and any positive function f continuous on $[r,t]$,

$$\widetilde{\int}_r^s f \cdot \widetilde{\int}_s^t f = \widetilde{\int}_r^t f.$$

2.7 THE FUNDAMENTAL THEOREMS OF GEOMETRIC CALCULUS

The geometric derivative and integral are 'inversely' related in the sense indicated by the following two theorems.

First Fundamental Theorem

If f is positive and continuous on [r,s], and

$$g(x) = \int_r^{\widetilde{x}} f, \quad \text{for every number x in } [r,s],$$

then

$$\widetilde{D}g = f, \quad \text{on } [r,s].$$

Second Fundamental Theorem

If $\widetilde{D}h$ is continuous on [r,s], then

$$\int_r^{\widetilde{s}} (\widetilde{D}h) = h(s)/h(r).$$

Just as the Second Fundamental Theorem of Classical Calculus is useful for evaluating classical integrals, so is the Second Fundamental Theorem of Geometric Calculus useful for evaluating geometric integrals. For example, let f(x) = exp(1/x) and h(x) = x, for x > 0. Then f = $\widetilde{D}h$, and so

$$\int_3^{\widetilde{5}} f = \int_3^{\widetilde{5}} (\widetilde{D}h) = h(5)/h(3) = 5/3.$$

2.8 RELATIONSHIPS TO THE CLASSICAL CALCULUS

Our presentation of the geometric calculus was independent of the classical calculus, for the operators of the former were not defined in terms of the operators of the latter. However, the corresponding operators of the two calculi are uniformly related.

Let f be a positive function and set $\bar{f}(x) = \ln(f(x))$. Let $G_r^s\bar{f}$ be the classical gradient of $\bar{f}$ over [r,s], and let $\widetilde{G}_r^s f$ be the geometric gradient of f over [r,s]. Then we have the following uniform relationships.

$$(1) \quad \widetilde{G}_r^s f = \exp\{G_r^s \bar{f}\}$$

$$(2) \quad [\widetilde{D}f](a) = \exp\{[D\bar{f}](a)\}$$

$$(3) \quad \widetilde{M}_r^s f = \exp\{M_r^s \bar{f}\}$$

$$(4) \quad \int_r^{\widetilde{s}} f = \exp\left\{\int_r^s \bar{f}\right\}$$

Remark. For (2) we assume that $[\widetilde{D}f](a)$ exists; for (3) and (4), that f is continuous on [r,s].

The preceding observations clearly indicate that for each theorem in classical calculus there is a corresponding theorem in geometric calculus, and conversely. For example, we have the following mean value theorem of geometric calculus.

> If a positive function f is continuous on [r,s]
> and geometrically differentiable everywhere be-
> tween r and s, then between r and s there is a
> number at which the geometric derivative of f
> equals the geometric gradient of f over [r,s].

The relationship of the nth-geometric derivative to the nth-classical derivative follows the familiar pattern:

$$[\widetilde{D}^n f](a) = \exp\{[D^n \bar{f}](a)\},$$

where $\bar{f}(x) = \ln(f(x))$.

The second geometric derivative of the function $\exp(-x^2)$ turns out to be the constant $1/e^2$, a fact that may provide additional insight into that important function.

Chapter 3

THE ANAGEOMETRIC CALCULUS

3.1 INTRODUCTION

In the anageometric calculus, changes in function arguments and values are measured by ratios and differences, respectively, and the operators are applied only to functions whose arguments are positive. However, an anageometric-type calculus for functions with negative arguments is discussed briefly in Section 6.10.

A positive interval is an interval [r,s] for which $0 < r < s$. The geometric extent of a positive interval [r,s] is the number s/r.

We were interested to learn that in his Principia, Newton expressed Galileo's law of descent of freely falling bodies as follows: "When a body is falling,...the spaces described in proportional times are as...the squares of the times." (Scholium to Corollary VI in "Axioms, or Laws of Motion.") That is, in time intervals of equal geometric extent, the distances traversed are proportional to the squares of the times elapsed.

Heuristic guides for selecting appropriate gradients, derivatives, averages, and integrals are discussed in Chapter 9.

3.2 THE ANAGEOMETRIC GRADIENT

An anageometrically-uniform function is a function that is on R_+, is continuous, and has the same classical change over any two positive intervals of equal geometric extent. Clearly every constant function on R_+ is anageometrically-uniform.

The anageometrically-uniform functions are those expressible in the form $\ln(cx^m)$, where c and m are constants, $c > 0$, and x is unrestricted in R_+.

It is characteristic of an anageometrically-uniform function that for each geometric progression of arguments, the corresponding sequence of values is an arithmetic progression.

The anageometric slope of an anageometrically-uniform

18

function is its classical change over any positive interval
of geometric extent e. (In Note 3 and Section 6.10 we explain why e was chosen here.) The anageometric slope of the
function $\ln(cx^m)$ turns out to be m.

If one plots an anageometrically-uniform function h on
semi-log paper that is logarithmically scaled on the x-axis,
the result is a straight line whose classical slope equals
the anageometric slope of h. Other aspects of graphical interpretation will be discussed more fully in Section 6.11.

The anageometric gradient of a function f over a positive interval [r,s] is the anageometric slope of the anageometrically-uniform function containing (r,f(r)) and (s,f(s)),
and turns out to be

$$\frac{f(s) - f(r)}{\ln(s) - \ln(r)} .$$

It is easy to verify that the anageometric gradient is
independent of the unit and origin used for measuring the
magnitudes which the arguments and values may represent, respectively.

The foregoing expression for the anageometric gradient
yields the indeterminate form 0/0 when r = s.

3.3 THE ANAGEOMETRIC DERIVATIVE

Throughout this section, f is assumed to be a function
defined at least on a positive interval containing the number a in its interior.

If the following limit exists, we denote it by [Df](a),
call it the anageometric derivative of f at a, and say that
f is anageometrically differentiable at a:

$$\lim_{x \to a} \frac{f(x) - f(a)}{\ln(x) - \ln(a)} .$$

It can be proved that [Df](a) and [Df](a) coexist; that
if they do exist, then

$$[Df](a) = [Df](a)/[D(\ln)](a) = a \cdot [Df](a);$$

and that [Df](a) equals the anageometric slope of the unique anageometrically-uniform function which is classically tangent to f at $(a,f(a))$.

Observe that $[\underset{\sim}{D}f](a)$ equals the classical derivative of f with respect to ln at a.

The anageometric derivative of f, denoted by $\underset{\sim}{D}f$, is the function that assigns to each number t the number $[\underset{\sim}{D}f](t)$, if it exists.

The operator $\underset{\sim}{D}$ is additive and homogeneous:

$$\underset{\sim}{D}(f + g) = \underset{\sim}{D}f + \underset{\sim}{D}g,$$

$$\underset{\sim}{D}(c \cdot f) = c \cdot \underset{\sim}{D}f, \quad c \text{ constant.}$$

If f is anageometrically-uniform, then $\underset{\sim}{D}f$ has a constant value equal to the anageometric slope of f. Indeed, only anageometrically-uniform functions have anageometric derivatives that are constant on R_+.

It is worth noting that if $h(x) = mx$, where m is a constant and $x > 0$, then $\underset{\sim}{D}h = h$.

3.4 THE ANAGEOMETRIC AVERAGE

Since changes in function arguments are measured by ratios in the anageometric calculus, one should not be surprised that the definition of the anageometric average requires the use of partitions in which the successive points form a geometric progression.

A geometric partition of a positive interval $[r,s]$ is any geometric progression whose first and last terms are r and s, respectively. A geometric partition is n-fold if it has n terms.

The anageometric average of a continuous function f on a positive interval $[r,s]$ is denoted by $\underset{\sim}{M}{}_r^s f$ and defined to be the limit of the convergent sequence whose nth term is the arithmetic average of $f(a_1),\ldots,f(a_n)$, where $a_1,\ldots,a_n$ is the n-fold geometric partition of $[r,s]$.

The arithmetic and anageometric averages are NOT identical; for example, if $f(x) = x$, for $x > 0$, then

$$\underset{\sim}{M}{}_r^s f = (s - r)/[\ln(s) - \ln(r)],$$

whereas

$$M_r^s f = (s + r)/2.$$

Nevertheless, the operator $\underset{\sim}{M}{}_r^s$ is additive and homogeneous:

$$\underset{\sim}{M}{}_r^s(f + g) = \underset{\sim}{M}{}_r^s f + \underset{\sim}{M}{}_r^s g,$$

$$\underset{\sim}{M}{}_r^s(c \cdot f) = c \cdot \underset{\sim}{M}{}_r^s f, \quad c \text{ constant.}$$

The anageometric average of an anageometrically-uniform function on $[r,s]$ is equal to the arithmetic average of its values at r and s, and is equal to its value at the geometric average of r and s.

The anageometric average is the only operator that has the following three properties.

1. For any positive interval $[r,s]$ and any constant function $h(x) = b$ on $[r,s]$,

$$\underset{\sim}{M}{}_r^s h = b.$$

2. For any positive interval $[r,s]$ and any functions f and g that are continuous on $[r,s]$, if $f(x) \leq g(x)$ for every number x in $[r,s]$, then

$$\underset{\sim}{M}{}_r^s f \leq \underset{\sim}{M}{}_r^s g.$$

3. For any numbers r,s,t such that $0 < r < s < t$, and any function f continuous on $[r,t]$,

$$[\ln(s) - \ln(r)] \cdot \underset{\sim}{M}{}_r^s f + [\ln(t) - \ln(s)] \cdot \underset{\sim}{M}{}_s^t f$$

$$= [\ln(t) - \ln(r)] \cdot \underset{\sim}{M}{}_r^t f.$$

3.5 THE BASIC THEOREM OF ANAGEOMETRIC CALCULUS

In view of the following theorem we say that the anageometric average fits naturally into the scheme of anageometric calculus. (We forego the discrete analogue.)

The Basic Theorem of Anageometric Calculus

If $\underset{\sim}{D}h$ is continuous on a positive interval $[r,s]$, then its anageometric average on $[r,s]$ equals the anageometric gradient of h over $[r,s]$, that is,

$$\underset{\sim}{M}{}_r^s(\underset{\sim}{D}h) = \frac{h(s) - h(r)}{\ln(s) - \ln(r)}.$$

The preceding theorem provides an immediate solution to

the following problem, which will motivate our definition of
the anageometric integral.

<u>The Basic Problem of Anageometric Calculus</u>

> Suppose that the value of a function h is known
> at a positive argument r, and suppose that f,
> the anageometric derivative of h, is continuous
> and known at each number in [r,s]. Find h(s).

<u>Solution</u>

$$h(s) = h(r) + [\ln(s) - \ln(r)] \cdot \underset{\sim r}{M^s} f.$$

3.6 THE ANAGEOMETRIC INTEGRAL

The <u>anageometric integral</u> of a continuous function f on
a positive interval [r,s] is the number $[\ln(s) - \ln(r)] \cdot \underset{\sim r}{M^s} f$
and is denoted by $\displaystyle\int_{\sim r}^{s} f$. We set $\displaystyle\int_{\sim r}^{r} f = 0$.

The anageometric integral is a weighted anageometric
average, since

$$\int_{\sim r}^{s} f = \underset{\sim r}{M^s}\{[\ln(s) - \ln(r)] \cdot f\}.$$

Furthermore, $\displaystyle\int_{\sim r}^{s} f$ equals the limit of the convergent sequence whose nth term is the sum

$$[\ln(k_n)] \cdot f(a_1) + \ldots + [\ln(k_n)] \cdot f(a_{n-1}),$$

where $a_1, \ldots, a_n$ is the n-fold geometric partition of [r,s],
and k_n is the common value of a_2/a_1, $a_3/a_2, \ldots, a_n/a_{n-1}$.

The reader who is familiar with Stieltjes integrals
will notice that the preceding result justifies the asser-
tion that

$$\int_{\sim r}^{s} f$$

is the Stieltjes integral of f with respect to ln on [r,s].

The operator $\displaystyle\int_{\sim r}^{s}$ is additive and homogeneous:

$$\int_{\sim r}^{s} (f + g) = \int_{\sim r}^{s} f + \int_{\sim r}^{s} g,$$

$$\int_{\sim r}^{s} (c \cdot f) = c \cdot \int_{\sim r}^{s} f, \quad c \text{ constant.}$$

The anageometric integral is the only operator that has the following three properties.

1. For any positive interval [r,s] and any constant function h(x) = b on [r,s],

$$\int_{\sim r}^{s} h = [\ln(s) - \ln(r)] \cdot b.$$

2. For any positive interval [r,s] and any functions f and g that are continuous on [r,s], if f(x) $\leq$ g(x) for every number x in [r,s], then

$$\int_{\sim r}^{s} f \leq \int_{\sim r}^{s} g.$$

3. For any numbers r,s,t such that 0 < r < s < t, and any function f continuous on [r,t],

$$\int_{\sim r}^{s} f + \int_{\sim s}^{t} f = \int_{\sim r}^{t} f.$$

3.7 THE FUNDAMENTAL THEOREMS OF ANAGEOMETRIC CALCULUS

The anageometric derivative and integral are 'inversely' related in the sense indicated by the following two theorems.

First Fundamental Theorem

If f is continuous on a positive interval [r,s], and

$$g(x) = \int_{\sim r}^{x} f, \quad \text{for every number x in [r,s],}$$

then

$$\underset{\sim}{D}g = f, \quad \text{on [r,s].}$$

Second Fundamental Theorem

If $\underset{\sim}{D}h$ is continuous on a positive interval [r,s], then

$$\int_{\sim r}^{s} (\underset{\sim}{D}h) = h(s) - h(r).$$

The preceding theorem is useful, of course, for evaluating anageometric integrals.

3.8 RELATIONSHIPS TO THE CLASSICAL CALCULUS

Let $\underset{\sim}{G}{}_r^s f$ be the anageometric gradient of f over a positive interval [r,s], and set $\bar{f}(x) = f(e^x)$, $\bar{r} = \ln(r)$, $\bar{s} = \ln(s)$, and $\bar{a} = \ln(a)$. (Of course, $G_{\bar{r}}^{\bar{s}}\bar{f}$ is the classical gradient of $\bar{f}$ over $[\bar{r},\bar{s}]$.) Then we have the following uniform relationships between the corresponding operators of the anageometric and classical calculi.

$$(1) \qquad \underset{\sim}{G}{}_r^s f = G_{\bar{r}}^{\bar{s}}\bar{f}$$

$$(2) \qquad [\underset{\sim}{D}f](a) = [D\bar{f}](\bar{a})$$

$$(3) \qquad \underset{\sim}{M}{}_r^s f = M_{\bar{r}}^{\bar{s}}\bar{f}$$

$$(4) \qquad \underset{\sim}{\int}_r^s f = \int_{\bar{r}}^{\bar{s}} \bar{f}$$

Remark. For (2), we assume that $[\underset{\sim}{D}f](a)$ exists; for (3) and (4), that f is continuous on [r,s].

The preceding observations clearly indicate that for each theorem in classical calculus there is a corresponding theorem in anageometric calculus, and conversely.

Chapter 4
THE BIGEOMETRIC CALCULUS

4.1 INTRODUCTION

In the bigeometric calculus, changes in function arguments and values are measured by ratios, and the operators are applied only to functions with positive arguments and positive values. However, bigeometric-type calculi for other functions can be constructed (Section 6.10).

Heuristic guides for selecting appropriate gradients, derivatives, averages, and integrals are discussed in Chapter 9.

4.2 THE BIGEOMETRIC GRADIENT

A <u>bigeometrically-uniform</u> <u>function</u> is a positive function that is on R_+, is continuous, and has the same geometric change over any two positive intervals of equal geometric extent. Clearly every positive constant function on R_+ is bigeometrically-uniform.

The bigeometrically-uniform functions are those expressible in the form cx^m, where c and m are constants, $c > 0$, and x is unrestricted in R_+.

It is characteristic of a bigeometrically-uniform function that for each geometric progression of arguments, the corresponding sequence of values is also a geometric progression.

The <u>bigeometric</u> <u>slope</u> of a bigeometrically-uniform function is its geometric change over any positive interval of geometric extent e. (Our reason for choosing e here is similar to the reason for using e in defining anageometric slope, as explained in Note 3 and Section 6.10.) The bigeometric slope of the function cx^m turns out to be e^m.

If one plots a bigeometrically-uniform function h on log-log paper, the result is a straight line whose classical slope equals the natural logarithm of the bigeometric slope of h. Other aspects of graphical interpretation will be discussed more fully in Section 6.11.

The <u>bigeometric</u> <u>gradient</u> of a positive function f over a

25

positive interval [r,s] is the bigeometric slope of the bigeometrically-uniform function containing (r,f(r)) and
(s,f(s)), and turns out to be

$$\left[\frac{f(s)}{f(r)}\right]^{1/[\ln(s)-\ln(r)]} \quad .$$

It is easy to verify that the bigeometric gradient is
independent of the units used for measuring the magnitudes
which the arguments and values may represent.

The foregoing expression for the bigeometric gradient
yields the indeterminate form 1^{∞} when $r = s$.

4.3 THE BIGEOMETRIC DERIVATIVE

Throughout this section, f is assumed to be a positive
function defined at least on a positive interval containing
the number a in its interior.

If the following limit exists and is positive, we de-
note it by $[\tilde{D}f](a)$, call it the __bigeometric derivative__ of f
at a, and say that f is bigeometrically differentiable at a:

$$\lim_{x \to a} \left[\frac{f(x)}{f(a)}\right]^{1/[\ln(x)-\ln(a)]} \quad .$$

Our reason for requiring $[\tilde{D}f](a)$ to be positive will be indicated in Section 6.10.

It can be proved that $[Df](a)$ and $[\tilde{D}f](a)$ coexist; that
if they do exist, then

$$[\tilde{D}f](a) = \exp\{a \cdot [Df](a)/f(a)\};$$

and that $[\tilde{D}f](a)$ equals the bigeometric slope of the unique

bigeometrically-uniform function which is classically tangent to f at (a,f(a)).

The bigeometric derivative of f, denoted by $\tilde{D}f$, is the
function that assigns to each number t the number $[\tilde{D}f](t)$,
if it exists.

The operator $\tilde{D}$ is multiplicative:

$$\tilde{D}(f \cdot g) = \tilde{D}f \cdot \tilde{D}g.$$

Furthermore,

$$\tilde{D}(f^{c}) = (\tilde{D}f)^{c}, \quad c \text{ constant.}$$

If f is bigeometrically-uniform, then $\tilde{\mathrm{D}}$f has a constant value equal to the bigeometric slope of f. Indeed, only bigeometrically-uniform functions have bigeometric derivatives that are constant on R_+.

It is also worth noting that if h(x) = exp(x) for x > 0, then $\tilde{\mathrm{D}}$h = h.

We noted above that

$$[\tilde{\mathrm{D}}f](a) = \exp\{a\cdot[Df](a)/f(a)\}.$$

Since economists refer to the expression within the braces as the elasticity of f at a, we call $[\tilde{\mathrm{D}}f](a)$ the <u>resiliency</u> of f at a. We believe that resiliency will prove to be more useful than elasticity because the former is the derivative in a complete system of calculus (the naturalness of which is exhibited in this chapter), whereas it appears to be impossible to construct a complete, natural system of calculus in which the derivative is the elasticity.

Perhaps the psychophysicists will find some interest in the bigeometric calculus, for one of their basic laws may be stated thus: The resiliency of the stimulus-sensation function is constant. (That constant is determined by the nature of the stimulus.)

The bigeometric calculus may also prove to be useful in biology, for a fundamental law of growth is the following: If f is the function relating the size of one organ to the size of any other given organ in the same body at the same instant, then, within certain time limits, the resiliency of f is constant.

A physicist who preferred not to settle on specific units of time and distance could, nevertheless, assert that the bigeometric speed of an object falling freely to the earth is constant.

Heuristic guides for selecting appropriate derivatives are discussed more fully in Section 9.2.

4.4 THE BIGEOMETRIC AVERAGE

The <u>bigeometric</u> <u>average</u> of a continuous positive function f on a positive interval [r,s] is denoted by $\tilde{\mathrm{M}}_r^s f$ and defined to be the positive limit of the convergent sequence whose nth term is the geometric average of $f(a_1),\ldots,f(a_n)$, where $a_1,\ldots,a_n$ is the n-fold geometric partition of $[r,s]$.

The geometric and bigeometric averages are not identical; for example, if $f(x) = x^2$ for $x > 0$, then

$$\widetilde{\underset{\sim}{M}}_1^e f = \exp\{2/(e - 1)\},$$

whereas

$$\widetilde{\underset{\sim}{M}}_1^e f = e.$$

Nevertheless, the operator $\widetilde{\underset{\sim}{M}}_r^s$ is multiplicative:

$$\widetilde{\underset{\sim}{M}}_r^s(f \cdot g) = \widetilde{\underset{\sim}{M}}_r^s f \cdot \widetilde{\underset{\sim}{M}}_r^s g.$$

Also,

$$\widetilde{\underset{\sim}{M}}_r^s(f^c) = (\widetilde{\underset{\sim}{M}}_r^s f)^c, \quad c \text{ constant}.$$

The bigeometric average of a bigeometrically-uniform function on $[r,s]$ is equal to the geometric average of its values at r and s, and is equal to its value at the geometric average of r and s.

The bigeometric average is the only operator that has the following three properties.

1. For any positive interval $[r,s]$ and any positive constant function $h(x) = b$ on $[r,s]$,

$$\widetilde{\underset{\sim}{M}}_r^s h = b.$$

2. For any positive interval $[r,s]$ and any positive functions f and g that are continuous on $[r,s]$, if $f(x) \le g(x)$ for every number x in $[r,s]$, then

$$\widetilde{\underset{\sim}{M}}_r^s f \le \widetilde{\underset{\sim}{M}}_r^s g.$$

3. For any numbers r,s,t such that $0 < r < s < t$, and any positive function f continuous on $[r,t]$,

$$[\widetilde{\underset{\sim}{M}}_r^s f]^{[\ln(s)-\ln(r)]} \cdot [\widetilde{\underset{\sim}{M}}_s^t f]^{[\ln(t)-\ln(s)]}$$
$$= [\widetilde{\underset{\sim}{M}}_r^t f]^{[\ln(t)-\ln(r)]}.$$

4.5 THE BASIC THEOREM OF BIGEOMETRIC CALCULUS

In view of the following theorem we say that the bigeometric average fits naturally into the scheme of bigeometric calculus. (We forego the discrete analogue.)

The Basic Theorem of Bigeometric Calculus

If $\tilde{D}h$ is continuous on a positive interval $[r,s]$, then its bigeometric average on $[r,s]$ equals the bigeometric gradient of h over $[r,s]$, that is,

$$\tilde{M}{}_{\tilde{r}}^{s}(\tilde{D}h) = \left[\frac{h(s)}{h(r)}\right]^{1/[\ln(s)-\ln(r)]}$$

The preceding theorem provides an immediate solution to the following problem, which will motivate our definition of the bigeometric integral.

The Basic Problem of Bigeometric Calculus

Suppose that the value of a positive function h is known at a positive argument r, and suppose that f, the bigeometric derivative of h, is continuous and known at each number in $[r,s]$. Find $h(s)$.

Solution

$$h(s) = h(r) \cdot [\tilde{M}{}_{\tilde{r}}^{s}f]^{[\ln(s)-\ln(r)]}.$$

4.6 THE BIGEOMETRIC INTEGRAL

The <u>bigeometric</u> <u>integral</u> of a continuous positive function f on a positive interval $[r,s]$ is the positive number

$$[\tilde{M}{}_{\tilde{r}}^{s}f]^{[\ln(s)-\ln(r)]}$$

and is denoted by $\displaystyle\int_{\tilde{r}}^{\tilde{s}} f$. We set $\displaystyle\int_{\tilde{r}}^{\tilde{r}} f = 1$.

The bigeometric integral is a weighted bigeometric average, since

$$\int_{\tilde{r}}^{\tilde{s}} f = \tilde{M}{}_{\tilde{r}}^{s}\left\{f^{[\ln(s)-\ln(r)]}\right\}.$$

Furthermore, $\displaystyle\int_{\tilde{r}}^{\tilde{s}} f$ equals the positive limit of the convergent sequence whose nth term is the product

$$[f(a_1)]^{\ln(k_n)} \cdot [f(a_2)]^{\ln(k_n)} \cdots [f(a_{n-1})]^{\ln(k_n)},$$

where $a_1, \ldots, a_n$ is the n-fold geometric partition of $[r,s]$, and k_n is the common value of a_2/a_1, a_3/a_2, \ldots, a_n/a_{n-1}.

The operator $\displaystyle\int_{\sim r}^{\sim s}$ is multiplicative:

$$\int_{\sim r}^{\sim s} (f \cdot g) = \int_{\sim r}^{\sim s} f \cdot \int_{\sim r}^{\sim s} g.$$

Also,

$$\int_{\sim r}^{\sim s} (f^c) = \left(\int_{\sim r}^{\sim s} f \right)^c, \quad c \text{ constant.}$$

The bigeometric integral is the only operator that has the following three properties.

1. For any positive interval $[r,s]$ and any positive constant function $h(x) = b$ on $[r,s]$,

$$\int_{\sim r}^{\sim s} h = b^{\ln(s) - \ln(r)}.$$

2. For any positive interval $[r,s]$ and any positive functions f and g that are continuous on $[r,s]$, if $f(x) \leq g(x)$ for every number x in $[r,s]$, then

$$\int_{\sim r}^{\sim s} f \leq \int_{\sim r}^{\sim s} g.$$

3. For any numbers r,s,t such that $0 < r < s < t$, and any positive function f continuous on $[r,t]$,

$$\int_{\sim r}^{\sim s} f \cdot \int_{\sim s}^{\sim t} f = \int_{\sim r}^{\sim t} f.$$

4.7 THE FUNDAMENTAL THEOREMS OF BIGEOMETRIC CALCULUS

The bigeometric derivative and integral are 'inversely' related in the sense indicated by the following two theorems.

First Fundamental Theorem

If f is positive and continuous on a positive interval $[r,s]$, and

$$g(x) = \int_{\sim r}^{\sim x} f, \text{ for every number } x \text{ in } [r,s],$$

then

$$\underset{\sim}{D} g = f, \quad \text{on } [r,s].$$

Second Fundamental Theorem

If $\underset{\sim}{D}h$ is continuous on a positive interval $[r,s]$, then

$$\int_{\underset{\sim}{r}}^{\tilde{s}} (\underset{\sim}{D}h) = h(s)/h(r).$$

The preceding theorem is useful for evaluating bigeometric integrals.

4.8 RELATIONSHIPS TO THE CLASSICAL CALCULUS

Let $\underset{\underset{\sim}{r}}{\overset{\tilde{s}}{G}}f$ be the bigeometric gradient of a positive function f over a positive interval $[r,s]$, and set $\bar{f}(x) = \ln(f(e^x))$, $\bar{r} = \ln(r)$, $\bar{s} = \ln(s)$, and $\bar{a} = \ln(a)$. (Of course, $\underset{\bar{r}}{\overset{\bar{s}}{G}}\bar{f}$ is the classical gradient of $\bar{f}$ over $[\bar{r},\bar{s}]$.) Then we have the following uniform relationships between the corresponding operators of the bigeometric and classical calculi.

$$(1) \qquad \underset{\underset{\sim}{r}}{\overset{\tilde{s}}{G}}f = \exp\{\underset{\bar{r}}{\overset{\bar{s}}{G}}\bar{f}\}$$

$$(2) \qquad [\underset{\sim}{D}f](a) = \exp\{[D\bar{f}](\bar{a})\}$$

$$(3) \qquad \underset{\underset{\sim}{r}}{\overset{\tilde{s}}{M}}f = \exp\{\underset{\bar{r}}{\overset{\bar{s}}{M}}\bar{f}\}$$

$$(4) \qquad \int_{\underset{\sim}{r}}^{\tilde{s}} f = \exp\left\{\int_{\bar{r}}^{\bar{s}} \bar{f}\right\}$$

Remark. For (2) we assume that $[\underset{\sim}{D}f](a)$ exists; for (3) and (4), that f is continuous on $[r,s]$.

The preceding observations clearly indicate that for each theorem in classical calculus there is a corresponding theorem in bigeometric calculus, and conversely.

Chapter 5

SYSTEMS OF ARITHMETIC

5.1 INTRODUCTION

In this chapter we discuss the general concept of an arithmetic, without which it would be impractical to construct the non-Newtonian calculi in Chapter 6. We also present here one specific arithmetic, the geometric arithmetic, which will be used in Section 6.10. In Chapter 7 we shall discuss and use the quadratic arithmetic, and in Chapter 8, the harmonic arithmetic.

5.2 ARITHMETICS

The concept of a complete ordered field evolved from the axiomatization of the real number system, whose basic ideas are assumed known to the reader.[*] Informally, a complete ordered field is a system consisting of a set A, four (binary) operations, $\dot{+}$, $\dot{-}$, $\dot{\times}$, $\dot{/}$ for A, and an ordering relation $\dot{<}$ for A, all of which behave with respect to A exactly as $+$, $-$, $\times$, $/$, $<$ behave with respect to the set of all numbers. We call A the realm of the complete ordered field.

There are infinitely-many complete ordered fields, all of which are structurally equivalent (isomorphic). The real number system is nowadays conceived as an arbitrarily selected complete ordered field whose realm is denoted by R and whose operations and ordering relation are denoted by $+$, $-$, $\times$, $/$, and $<$.

By an arithmetic we mean a complete ordered field whose realm is a subset of R. (Although the term "arithmetic" is commonly used for referring to the system or study of the positive integers, we have taken the liberty of using the term in the sense just indicated.) There are infinitely-many arithmetics, one of which is the real num-

---

[*] Complete ordered fields are discussed in many textbooks on modern algebra and in some works on analysis. A detailed, new axiomatic treatment of the real number system is presented in the authors' Axiomatic Analysis, obtainable from Lee Press.

32

ber system, henceforth called the <u>classical</u> <u>arithmetic</u>.

The rules for handling any arithmetic $(A, \dot{+}, \dot{-}, \dot{\times}, \dot{/}, \dot{<})$ are exactly the same as the rules for handling classical arithmetic. For example, $\dot{+}$ and $\dot{\times}$ are commutative and associative; $\dot{\times}$ is distributive with respect to $\dot{+}$; $\dot{<}$ is transitive; and there are two unique numbers $\dot{0}$ and $\dot{1}$ in A such that $y \dot{+} \dot{0} = y$ and $y \dot{\times} \dot{1} = y$, for every number y in A. The notation "$y \dot{\leq} z$" is, of course, an abbreviation of "$y \dot{<} z$ or $y = z$."

Although all arithmetics are structurally equivalent, only by distinguishing among them do we obtain suitable tools for constructing all the non-Newtonian calculi. But the usefulness of arithmetics is not limited to the construction of calculi; we believe there is a more fundamental reason for considering alternative arithmetics: they may also be helpful in developing and understanding new systems of measurement that could yield simpler physical laws.

In his <u>Foundations</u> <u>of</u> <u>Science</u> (formerly titled, <u>Physics:</u> <u>The</u> <u>Elements</u>), Norman Robert Campbell, a pioneer in the theory of measurement, clearly recognized that alternative arithmetics might be useful in science, for he wrote, "we must recognize the possibility that a system of measurement may be arbitrary otherwise than in the choice of unit; there may be arbitrariness in the choice of the process of addition."

5.3 α-ARITHMETIC

A <u>generator</u> is a one-to-one function whose domain is R and whose range is a subset of R. For example, the following are generators: the identity function I, the function exp, and the function x^3.

Consider any generator α with range A. By <u>α-arithmetic</u> we mean the arithmetic whose realm is A and whose operations and ordering relation are defined as follows.

$$\alpha\text{-addition}\ldots\ldots\ldots\ldots\ y \mathbin{\dot{+}} z = \alpha\{\alpha^{-1}(y) + \alpha^{-1}(z)\}$$

$$\alpha\text{-subtraction}\ldots\ldots\ldots\ y \mathbin{\dot{-}} z = \alpha\{\alpha^{-1}(y) - \alpha^{-1}(z)\}$$

$$\alpha\text{-multiplication}\ldots\ldots\ y \mathbin{\dot{\times}} z = \alpha\{\alpha^{-1}(y) \times \alpha^{-1}(z)\}$$

$$\alpha\text{-division}\ (z \neq \dot{0})\ldots\ y \mathbin{\dot{/}} z = \alpha\{\alpha^{-1}(y) / \alpha^{-1}(z)\}$$

$$\alpha\text{-order}\ldots\ldots\ldots\ldots\ y \mathbin{\dot{<}} z \text{ if and only if}$$
$$\alpha^{-1}(y) < \alpha^{-1}(z)$$

We say that α _generates_ α-arithmetic; for example, the identity function generates classical arithmetic, and the function exp generates geometric arithmetic, which is discussed in Section 5.4. Each generator generates exactly one arithmetic, and, conversely, each arithmetic is generated by exactly one generator.

All concepts in classical arithmetic have natural counterparts in α-arithmetic, some of which we now discuss.

The α-_positive_ numbers are the numbers x in A such that $\dot{0} \mathbin{\dot{<}} x$; the α-_negative_ numbers are those for which $x \mathbin{\dot{<}} \dot{0}$. The α-_zero_, $\dot{0}$, and the α-_one_, $\dot{1}$, turn out to be $\alpha(0)$ and $\alpha(1)$. The α-_integers_ consist of $\dot{0}$ and all the numbers that result by successive α-addition of $\dot{1}$ to $\dot{0}$ and by successive α-subtraction of $\dot{1}$ from $\dot{0}$. Thus the α-integers turn out to be the following:

$$\ldots,\ \alpha(-2),\ \alpha(-1),\ \alpha(0),\ \alpha(1),\ \alpha(2),\ldots$$

For each integer n, we set $\dot{n} = \alpha(n)$. Of course, if $\dot{n}$ is an α-positive integer, then

$$\dot{n} = \underbrace{\dot{1} \mathbin{\dot{+}} \ldots \mathbin{\dot{+}} \dot{1}}_{n \text{ terms}}.$$

The α-_absolute_ value of a number x in A is

$$\dot{|x|} = \begin{cases} x & \text{if } x \mathbin{\dot{>}} \dot{0} \\ \dot{0} & \text{if } x = \dot{0} \\ \dot{0}\mathbin{\dot{-}}x & \text{if } x \mathbin{\dot{<}} \dot{0} \end{cases}$$

The $\underline{\alpha\text{-average}}$ of n numbers $u_1,\ldots,u_n$ in A is the unique
number u in A such that

$$u \dotplus \ldots \dotplus u = u_1 \dotplus \ldots \dotplus u_n.$$
$$\underbrace{\qquad\qquad\qquad}_{\text{n terms}}$$

It turns out that u equals

$$(u_1 \dotplus \ldots \dotplus u_n)\dotslash n,$$

which equals

$$\alpha\{[\alpha^{-1}(u_1) + \ldots + \alpha^{-1}(u_n)]/n\}.$$

We shall occasionally refer to the α-average as the $\underline{\text{nat-}}$
$\underline{\text{ural average}}$ in α-arithmetic. The natural average in classical arithmetic is the arithmetic average; the natural aver-
age in geometric arithmetic is the geometric average.

Although the α-average is widely known, we have seen no
reference to it as the natural counterpart in α-arithmetic of
the arithmetic average in classical arithmetic. Indeed, the
α-average has the same properties in the context of α-arithmetic as the arithmetic average has in the context of classical arithmetic. (An important example will be given in Section 10.4.)

An $\underline{\alpha\text{-progression}}$ (or $\underline{\text{natural progression}}$ in α-arithmetic)
is a finite sequence of numbers $u_1,\ldots,u_n$ in A such that
$u_{i+1} \dotminus u_i$ is the same for every integer i from 1 to n-1. In
classical arithmetic the natural progressions are the arithmetic progressions; in geometric arithmetic they are the geometric progressions.

For any numbers r and s in A, if $r \overset{\cdot}{<} s$, then the set of
all numbers x in A such that $r \overset{\cdot}{\leq} x \overset{\cdot}{\leq} s$ is called an $\underline{\alpha\text{-inter-}}$
$\underline{\text{val}}$, is denoted by $[r,s]$, has $\underline{\alpha\text{-extent}}$ of $s \dotminus r$, and has an
$\underline{\alpha\text{-interior}}$ consisting of all numbers x in A such that $r \overset{\cdot}{<}$
$x \overset{\cdot}{<} s$.

An $\underline{\alpha\text{-partition}}$ of an α-interval $[r,s]$ is any α-progression whose first and last terms are r and s. An α-partition
is $\underline{\text{n-fold}}$ if it has n terms.

Let $\{u_n\}$ be an infinite sequence of numbers in A. Then
there is at most one number u in A such that every α-interval
with u in its α-interior contains all but finitely-many terms

of $\{u_n\}$. If there is such a number u, then $\{u_n\}$ is said to be α-convergent to u, which is called the α-limit of $\{u_n\}$.

For $\alpha = I$, α-convergence reduces to classical convergence.

5.4 GEOMETRIC ARITHMETIC

The arithmetic generated by the function exp will be called geometric arithmetic rather than exp-arithmetic. Similarly, the notions in geometric arithmetic will be indicated by the adjective "geometric" rather than by the prefix "exp." For example, the natural average will be referred to as the geometric average, a usage that is consistent with generally accepted terminology.

In Section 6.10 we shall show that by apt use of geometric arithmetic one can readily obtain the geometric calculus, the anageometric calculus, and the bigeometric calculus.

Geometric arithmetic has the following features. (The letters y and z represent arbitrary positive numbers.)

Generator................. exp

Realm..................... R_+

Geometric zero........... 1   [= exp(0)]

Geometric one............ e   [= exp(1)]

Geometric sum .... $\exp\{\ln(y) + \ln(z)\}$
  of y and z $= yz$

Geometric difference ...... $\exp\{\ln(y) - \ln(z)\}$
  between y and z $= y/z$

Geometric product .... $\exp\{\ln(y) \cdot \ln(z)\}$
  of y and z $= y^{\ln(z)} = z^{\ln(y)}$

Geometric quotient ...... $\exp\{\ln(y)/\ln(z)\}$
  of y and z, $(z \neq 1)$ $= y^{1/\ln(z)}$

Geometric order.......... Identical with classical order

Geometric positive numbers Numbers greater than 1

Geometric negative numbers Positive numbers less than 1

Geometric intervals....... Identical with
                                positive intervals

Geometric extent of [y,z]..... z/y

Natural average............... Geometric average

Natural progressions.......... Geometric progressions

Geometric partition of [y,z].. Any geometric progression
                               whose first and last
                               terms are y and z

Geometric convergence is equivalent to classical convergence in the sense that a sequence $\{p_n\}$ of positive numbers geometrically converges to a positive number p if and only if $\{p_n\}$ classically converges to p.

Geometric arithmetic should be especially useful in situations where products and ratios provide the natural methods of combining and comparing magnitudes. Of course, geometric arithmetic applies only to positive numbers, since the range of exp is R_+. But it is a simple matter to construct a geometric-type arithmetic that applies to negative numbers: one simply uses the generator -exp, which assigns to each number x the negative number $-e^x$.

Chapter 6

THE *-CALCULUS

6.1 INTRODUCTION

For the remainder of this book, α and β are arbitrarily selected generators and * ("star") is the ordered pair of arithmetics (α-arithmetic, β-arithmetic). The following notations will be used.

	α-Arithmetic	β-Arithmetic
Realm.............	A	B
Addition.........	$\dot{+}$	$\ddot{+}$
Subtraction......	$\dot{-}$	$\ddot{-}$
Multiplication...	$\dot{\times}$	$\ddot{\times}$
Division.........	$\dot{/}$	$\ddot{/}$
Order...........	$\dot{<}$	$\ddot{<}$

It should be understood that all the definitions in Chapter 5 apply equally well to β-arithmetic. For example, β-convergence is defined by means of β-intervals and their β-interiors.

In the *-calculus, α-arithmetic is used on arguments and β-arithmetic is used on values; in particular, changes in arguments and values are measured by α-differences and β-differences, respectively. The operators of the *-calculus are applied only to functions with arguments in A and values in B. Accordingly, unless indicated or implied otherwise, all functions are assumed to be of that character.

Heuristic guides for selecting appropriate gradients, derivatives, integrals, and averages are discussed in Chapter 9.

The *-limit of a function f at a number a in A is, if it exists, the unique number b in B such that for every infinite sequence $\{a_n\}$ of arguments of f distinct from a, if $\{a_n\}$ α-converges to a, then $\{f(a_n)\}$ β-converges to b. We write

$$\text{*-}\lim_{x \to a} f(x) = b.$$

38

A function f is *-<u>continuous</u> at a number a in A if and
only if a is an argument of f and

$$\text{*-lim}_{x \to a} f(x) = f(a).$$

When α and β are the identity function I, the concepts
of *-limit and *-continuity are identical with those of classical limit and classical continuity, but that is possible
even when α and β do not equal I.

The <u>isomorphism</u> from α-arithmetic to β-arithmetic is
the unique function ι (iota) that possesses the following
three properties.

1. ι is one-to-one.

2. ι is on A and onto B.

3.   For any numbers u and v in A,

$$\iota(u \mathbin{\dot{+}} v) = \iota(u) \mathbin{\ddot{+}} \iota(v),$$

$$\iota(u \mathbin{\dot{-}} v) = \iota(u) \mathbin{\ddot{-}} \iota(v),$$

$$\iota(u \mathbin{\dot{\times}} v) = \iota(u) \mathbin{\ddot{\times}} \iota(v),$$

$$\iota(u \mathbin{\dot{/}} v) = \iota(u) \mathbin{\ddot{/}} \iota(v), \quad v \neq \dot{0},$$

$$u \mathbin{\dot{<}} v \quad \text{if and only if}$$

$$\iota(u) \mathbin{\ddot{<}} \iota(v).$$

It turns out that $\iota(x) = \beta\{\alpha^{-1}(x)\}$ for every number x
in A, and that $\iota(\dot{n}) = \ddot{n}$ for every integer n.

Since, for example, $u \mathbin{\dot{+}} v = \iota^{-1}\{\iota(u) \mathbin{\ddot{+}} \iota(v)\}$, it should
be clear that any statement in α-arithmetic can readily be
transformed into a statement in β-arithmetic.

6.2 THE *-<u>GRADIENT</u>

The β-<u>change</u> of a function f over an α-interval $[r,s]$
is the number $f(s) \mathbin{\ddot{-}} f(r)$ in B.

A *-<u>uniform</u> <u>function</u> is a function that is on A, is *-
continuous, and has the same β-change over any two α-inter-
vals of equal α-extent. Clearly every constant function on
A is *-uniform.

The *-uniform functions are those expressible in the
form $\iota\{(m \mathbin{\dot{\times}} x) \mathbin{\dot{+}} c\}$, where m and c are constants in A and

x is unrestricted in A. By choosing m = 1̇ and c = 0̇, we see
that ι is *-uniform.

It is characteristic of a *-uniform function that for
each α-progression of arguments, the corresponding sequence
of values is a β-progression.

The *-slope of a *-uniform function is its β-change
over any α-interval of α-extent 1̇. For example, the *-slope
of the function ι{(m ×̇ x) +̇ c} turns out to be ι(m). In particular, the *-slope of ι equals 1̈, and the *-slope of a constant function on A equals 0̈.

The *-gradient of a function f over [r,ṡ] is the *-slope
of the *-uniform function containing (r,f(r)) and (s,f(s)),
and turns out to be

$$[f(s) \,\ddot{-}\, f(r)] \ddot{/} [ι(s) \,\ddot{-}\, ι(r)].$$

The preceding expression yields the indeterminate form
0̈/̈0̈ when r = s.

6.3 THE *-DERIVATIVE

Throughout this section, f is assumed to be a function
defined at least on an α-interval containing the number a in
its α-interior.

If the following *-limit exists, we denote it by [D̊f](a),
call it the *-derivative of f at a, and say that f is *-dif-
ferentiable at a:

$$\text{*-}\lim_{x \to a}\left\{[f(x) \,\ddot{-}\, f(a)] \ddot{/} [ι(x) \,\ddot{-}\, ι(a)]\right\}.$$

If it exists, [D̊f](a) is necessarily in B.

The *-derivative of f, denoted by D̊f, is the function
that assigns to each number t in A the number [D̊f](t), if it
exists.

The operator D̊ is β-additive and β-homogeneous, that is

$$D̊(f \,\ddot{+}\, g) = D̊f \,\ddot{+}\, D̊g,$$

$$D̊(c \,\ddot{×}\, f) = c \,\ddot{×}\, D̊f, \quad c \text{ constant in B.}$$

If f is *-uniform, then D̊f has a constant value equal
to the *-slope of f. Indeed, only *-uniform functions have
*-derivatives that are constant on A.

Now we generalize the concept of classical tangent defined in Section 1.3.

The *-tangent to a function f at the point (a,f(a)) is the unique *-uniform function g, if it exists, which possesses the following two properties:

1.   g contains (a,f(a)).

2.   For each *-uniform function h containing (a,f(a)) and distinct from g, there is an α-positive number p such that for every number x in [a ∸ p, a ∔ p] but distinct from a,

$$\left| g(x) \mathbin{\ddot{\smile}} f(x) \right| \mathbin{\ddot{<}} \left| h(x) \mathbin{\ddot{\smile}} f(x) \right|.$$

It can be proved that $[\overset{*}{D}f](a)$ exists if and only if f has a *-tangent at (a,f(a)), and that if $[\overset{*}{D}f](a)$ does exist, it equals the *-slope of that *-tangent.

We say that two functions are *-tangent at a common point if and only if they have the same *-tangent there.

The derivatives $[Df](a)$ and $[\overset{*}{D}f](a)$ do not necessarily coexist and are seldom equal; however, if the following exist,

$$[D(\alpha^{-1})](a), \quad [D\alpha](\alpha^{-1}(a)), \quad [D(\beta^{-1})](f(a)), \quad [D\beta]\{\beta^{-1}(f(a))\},$$

then $[Df](a)$ and $[\overset{*}{D}f](a)$ do coexist, and the *-tangent to f at (a,f(a)) is, if it exists, classically tangent there.

6.4 THE *-AVERAGE

The *-average of a *-continuous function f on $[r,s]$ is denoted by $\overset{*}{M}{}_{r}^{s}f$ and defined to be the β-limit of the β-convergent sequence whose nth term is the β-average of $f(a_1),\dots,$ $f(a_n)$, where $a_1,\dots,a_n$ is the n-fold α-partition of $[r,s]$.

The operator $\overset{*}{M}{}_{r}^{s}$ is β-additive and β-homogeneous (Section 6.3).

The *-average of a *-uniform function on $[r,s]$ is equal to the β-average of its values at r and s, and is equal to its value at the α-average of r and s.

The *-average is the only operator that has the following three properties.

1. For any α-interval $[r,s]$ and any constant function
    $h(x) = b$ on $[r,s]$,

$$\overset{*s}{\underset{r}{M}}h = b.$$

2. For any α-interval $[r,s]$ and any functions f and g
    that are *-continuous on $[r,s]$, if $f(x) \leq g(x)$ for
    every number x in $[r,s]$, then

$$\overset{*s}{\underset{r}{M}}f \;\dot{\leq}\; \overset{*s}{\underset{r}{M}}g.$$

3. For any numbers r,s,t in A such that $r < s < t$,
    and any function f *-continuous on $[r,t]$,

$$[\iota(s) \ominus \iota(r)] \;\times\; \overset{*s}{\underset{r}{M}}f \;\oplus\; [\iota(t) \ominus \iota(s)] \;\times\; \overset{*t}{\underset{s}{M}}f$$

$$= \; [\iota(t) \ominus \iota(r)] \;\times\; \overset{*t}{\underset{r}{M}}f.$$

6.5 THE BASIC THEOREM OF *-CALCULUS

We begin with the discrete analogue.

The Discrete Analogue of the
Basic Theorem of *-Calculus

Let h be a discrete function whose arguments $a_1,\dots,a_n$
are an α-partition of $[r,s]$. Then the β-average of
the following $n-1$ *-gradients is equal to the *-gradi-
ent of h over $[r,s]$:

$$[h(a_{i+1}) \ominus h(a_i)] \;\oslash\; [\iota(a_{i+1}) \ominus \iota(a_i)],$$

$$i = 1,\dots,n-1.$$

In view of the following theorem we say that the *-aver-
age fits naturally into the scheme of *-calculus.

The Basic Theorem of *-Calculus

If $\overset{*}{D}h$ is *-continuous on $[r,s]$, then its *-average on
$[r,s]$ equals the *-gradient of h over $[r,s]$, that is,

$$\overset{*s}{\underset{r}{M}}(\overset{*}{D}h) = [h(s) \ominus h(r)] \oslash [\iota(s) \ominus \iota(r)].$$

The preceding theorem provides an immediate solution to
the following problem, which will motivate our definition of
the *-integral.

The Basic Problem of *-Calculus

Suppose that the value of a function h is known at
an argument r, and suppose that f, the *-derivative
of h, is *-continuous and known at each number in
$[r,s]$. Find h(s).

Solution

$$h(s) = h(r) \mathbin{\ddot{+}} \{[\iota(s) \mathbin{\ddot{-}} \iota(r)] \mathbin{\ddot{\times}} \overset{*s}{M_r} f\}.$$

6.6 THE *-INTEGRAL

The *-integral of a *-continuous function f on $[r,s]$,
denoted by $\overset{*}{\int_r^s} f$, is the following number in B:

$$[\iota(s) \mathbin{\ddot{-}} \iota(r)] \mathbin{\ddot{\times}} \overset{*s}{M_r} f.$$

We set $\overset{*}{\int_r^r} f = \ddot{0}$.

The *-integral is a weighted *-average, since

$$\overset{*}{\int_r^s} f = \overset{*s}{M_r} \{[\iota(s) \mathbin{\ddot{-}} \iota(r)] \mathbin{\ddot{\times}} f\}.$$

Furthermore, $\overset{*}{\int_r^s} f$ equals the β-limit of the β-convergent

sequence whose nth term is

$$[\iota(k_n) \mathbin{\ddot{\times}} f(a_1)] \mathbin{\ddot{+}} \ldots \mathbin{\ddot{+}} [\iota(k_n) \mathbin{\ddot{\times}} f(a_{n-1})],$$

where $a_1,\ldots,a_n$ is the n-fold α-partition of $[r,s]$, and k_n
is the common value of $a_2 \mathbin{\dot{-}} a_1$, $a_3 \mathbin{\dot{-}} a_2,\ldots,a_n \mathbin{\dot{-}} a_{n-1}$.

(If α is classically continuous and β = I, then the
*-integral is a Stieltjes integral.)

The *-integral is β-additive and β-homogeneous (Section
6.3) and is the only operator that has the following three
properties.

1.   For any α-interval $[r,s]$ and any constant function
h(x) = b on $[r,s]$,

$$\overset{*}{\int_r^s} h = [\iota(s) \mathbin{\ddot{-}} \iota(r)] \mathbin{\ddot{\times}} b.$$

2. For any α-interval $[r,s]$ and any functions f and g that are *-continuous on $[r,s]$, if $f(x) \underset{\cdot\cdot}{\leq} g(x)$ for every number x in $[r,s]$, then

$$\int_r^s \!\!\!{}^* f \;\; \underset{\cdot\cdot}{\leq} \;\; \int_r^s \!\!\!{}^* g.$$

3. For any numbers r,s,t in A such that $r \stackrel{\cdot}{<} s \stackrel{\cdot}{<} t$, and any function f *-continuous on $[r,t]$,

$$\int_r^s \!\!\!{}^* f \;\; \stackrel{\cdot\cdot}{+} \;\; \int_s^t \!\!\!{}^* f \;=\; \int_r^t \!\!\!{}^* f.$$

6.7 THE FUNDAMENTAL THEOREMS OF *-CALCULUS

The *-derivative and *-integral are 'inversely' related in the sense indicated by the following two theorems.

First Fundamental Theorem

If f is *-continuous on $[r,s]$, and

$$g(x) \;=\; \int_r^x \!\!\!{}^* f, \quad \text{for every number x in } [r,s],$$

then

$$\overset{*}{D}g = f, \quad \text{on } [r,s].$$

Second Fundamental Theorem

If $\overset{*}{D}h$ is *-continuous on $[r,s]$, then

$$\int_r^s \!\!\!{}^* (\overset{*}{D}h) \;=\; h(s) \stackrel{\cdot\cdot}{-} h(r).$$

6.8 RELATIONSHIPS TO THE CLASSICAL CALCULUS

In this section we indicate the uniform relationships between the corresponding notions of the *-calculus and classical calculus.

For each number a in A, let $\bar{a} = \alpha^{-1}(a)$. Let f be a function with arguments in A and values in B, and set $\bar{f}(t) = \beta^{-1}\{f(\alpha(t))\}$.

Then *-$\lim_{x \to a} f(x)$ and $\lim_{t \to \bar{a}} \bar{f}(t)$ coexist, and if they do exist,

$$\text{*-}\lim_{x \to a} f(x) \;=\; \beta\left\{\lim_{t \to \bar{a}} \bar{f}(t)\right\}.$$

Furthermore, f is *-continuous at a if and only if $\bar{f}$ is classically continuous at $\bar{a}$.

If $\overset{*}{G}{}^{s}_{r}f$ is the *-gradient of f over $[r,s]$, then

$$\overset{*}{G}{}^{s}_{r}f = \beta\{G^{\bar{s}}_{\bar{r}}\bar{f}\},$$

where $G^{\bar{s}}_{\bar{r}}\bar{f}$ is the classical gradient of $\bar{f}$ over $[\bar{r},\bar{s}]$.

The derivatives $[\overset{*}{D}f](a)$ and $[D\bar{f}](\bar{a})$ coexist, and if they do exist,

$$[\overset{*}{D}f](a) = \beta\{[D\bar{f}](\bar{a})\}.$$

And if f is *-continuous on $[r,s]$, then

$$\overset{*}{M}{}^{s}_{r}f = \beta\{M^{\bar{s}}_{\bar{r}}\bar{f}\}, \text{ and}$$

$$\overset{*}{\int}_{r}^{s} f = \beta\left\{\int_{\bar{r}}^{\bar{s}}\bar{f}\right\}.$$

The following facts are also worth noting. Let a ε A and b ε B. If α and β are classically continuous at $\alpha^{-1}(a)$ and $\beta^{-1}(b)$, respectively, then

$$\text{*-lim}_{x \to a} f(x) = b \quad \text{if and only if} \quad \lim_{x \to a} f(x) = b.$$

If α and β are classically continuous at $\alpha^{-1}(a)$ and $\beta^{-1}(f(a))$, respectively, then f is *-continuous at a if and only if f is classically continuous at a.

6.9 RELATIONSHIPS BETWEEN ANY TWO CALCULI

In this section we indicate the uniform relationships between the corresponding notions of any two given calculi, the *-calculus and the $\bar{*}$-calculus.

For the *-calculus there are two generators, α and β, which generate arithmetics with realms A and B. For the $\bar{*}$-calculus there are two generators, $\bar{\alpha}$ and $\bar{\beta}$, which generate arithmetics with realms $\bar{A}$ and $\bar{B}$.

Since all arithmetics are isomorphic, there is a unique isomorphism T_1 from α-arithmetic to $\bar{\alpha}$-arithmetic and a unique isomorphism T_2 from β-arithmetic to $\bar{\beta}$-arithmetic. For each number x in A let $\bar{x}$ be the number $T_1(x)$ in $\bar{A}$, and for each number y in B let $\bar{y}$ be the number $T_2(y)$ in $\bar{B}$. For each function f with arguments in A and values in B, let $\bar{f}$ be the function consisting of all ordered pairs $(\bar{x},\bar{y})$, where (x,y) is an arbitrary ordered pair in f. Clearly $\bar{f}$ has arguments in $\bar{A}$ and values in $\bar{B}$.

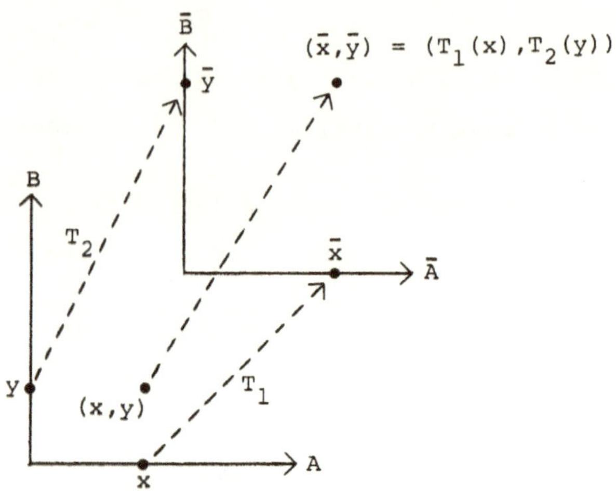

The following notations will also be used.

	*-Calculus	$\bar{*}$-Calculus
Limits.........	*-lim	$\bar{*}$-lim
Gradients......	$\overset{*}{G}$	$\overset{\bar{*}}{G}$
Derivatives....	$\overset{*}{D}$	$\overset{\bar{*}}{D}$
Averages.......	$\overset{*}{M}$	$\overset{\bar{*}}{M}$
Integrals......	$\overset{*}{\int}$	$\overset{\bar{*}}{\int}$

Let f be a function with arguments in A and values in B, and let the number a be in A. Then the following coexist:

$$\underset{x\to a}{\text{*-lim}} \, f(x), \quad \underset{t\to \bar{a}}{\bar{*}\text{-lim}} \, \bar{f}(t);$$

and if they do exist, then

$$\overline{\underset{x\to a}{\text{*-lim}} \, f(x)} = \underset{t\to \bar{a}}{\bar{*}\text{-lim}} \, \bar{f}(t).$$

Furthermore, f is *-continuous at a if and only if $\bar{f}$ is $\bar{*}$-continuous at $\bar{a}$.

If r and s are numbers in A such that $r \,\dot{<}\, s$, then

$$\overline{\overset{*s}{\underset{r}{G}}f} = \overset{\overline{*}\,\overline{s}}{\underset{\overline{r}}{G}}\overline{f}.$$

The derivatives $[\overset{*}{D}f](a)$ and $[\overline{\overset{\overline{*}}{D}\overline{f}}](\overline{a})$ coexist, and if they do exist,

$$\overline{[\overset{*}{D}f](a)} = [\overset{\overline{*}}{D}\overline{f}](\overline{a}).$$

Finally, if f is *-continuous on [r,s], then

$$\overline{\overset{*s}{\underset{r}{M}}f} = \overset{\overline{*}\,\overline{s}}{\underset{\overline{r}}{M}}\overline{f}, \quad \text{and}$$

$$\overline{\int_{r}^{*\,s} f} = \int_{\overline{r}}^{\overline{*}\,\overline{s}} \overline{f}.$$

The preceding observations clearly indicate that each theorem in any given calculus has an analogue, or correspondent, in every other calculus. In particular, each theorem in classical calculus has an analogue in *-calculus, and conversely. As an illustration we state the following theorem.

First Mean Value Theorem of *-Calculus

Let f be *-continuous on $[r,s]$ and *-differentiable at each number x in A for which $r \,\dot{<}\, x \,\dot{<}\, s$. Then there exists a number c in A such that $r \,\dot{<}\, c \,\dot{<}\, s$ and such that $[\overset{*}{D}f](c)$ equals the *-gradient of f over $[r,s]$.

6.10 APPLICATIONS OF GEOMETRIC ARITHMETIC

By appropriately specifying α and β one can obtain from the *-calculus the calculi developed in Chapters 1-4:

Calculus	α	β
classical.....	I	I
geometric.....	I	exp
anageometric..	exp	I
bigeometric...	exp	exp

Thus, one uses geometric arithmetic on function values in the geometric calculus, on function arguments in the anageometric calculus, and on both arguments and values in the bigeometric calculus.

In defining anageometric slope (Section 3.2), we used
geometric intervals of geometric extent e, one reason for
which appears in Note 3. But the fundamental reason stems
from our desire that the anageometric calculus be the *-cal-
culus for which α = exp and β = I, since there the *-slope
is defined by means of α-intervals of α-extent $\dot{1}$, which
equals e.

Similarly, geometric intervals of geometric extent e
were used in defining bigeometric slope (Section 4.2) to
assure that the bigeometric calculus be the *-calculus for
which α = β = exp.

Our definition of the geometric derivative (Section 2.3)
required that it be positive. The fundamental reason for the
restriction rests upon our desire that the geometric calculus
be the *-calculus for which α = I and β = exp, since there
the *-derivative,if it exists, must necessarily be in the
realm of β-arithmetic, that is, in R_+. A similar reason un-
derlies the stipulation that the bigeometric derivative be
positive (Section 4.3).

The operators of the geometric calculus are applied only
to functions with positive values. However, a geometric-type
calculus for negative (valued) functions can easily be obtained by choosing α = I and β = -exp.

By choosing α = -exp and β = I, one obtains an anageometric-type calculus for functions with negative arguments.
By choosing α = -exp and β = -exp, one gets a bigeometric-
type calculus for functions with negative arguments and values. Or one could, for example, choose α = -exp and β = exp,
thereby obtaining a bigeometric-type calculus for functions
with negative arguments and positive values.

Our emphasis, thus far, on the geometric family of calculi may be explained by the fact that these calculi possess
gradients and derivatives which are independent of the units
used in measuring certain magnitudes. The invariance of gradients and derivatives is discussed more generally in Section
9.2.

Of course an endless variety of calculi can be obtained
by using geometric arithmetic in tandem with other arithmetics.
For instance, one could choose α = exp and β = tanh. Other

specific calculi are treated in Chapters 7 and 8. In Note 4
we discuss briefly the sigmoidal arithmetic and related calculi, which may prove useful in statistics and biology.

6.11 GRAPHICAL INTERPRETATIONS

Except for the concept of *-tangency introduced in Section 6.3, we have thus far treated the *-calculus not geometrically but analytically, which was the way we conceived it.
In this section we give two graphical interpretations of the
*-calculus.

By *-paper we mean paper that is ruled off in squares
and marked thus:

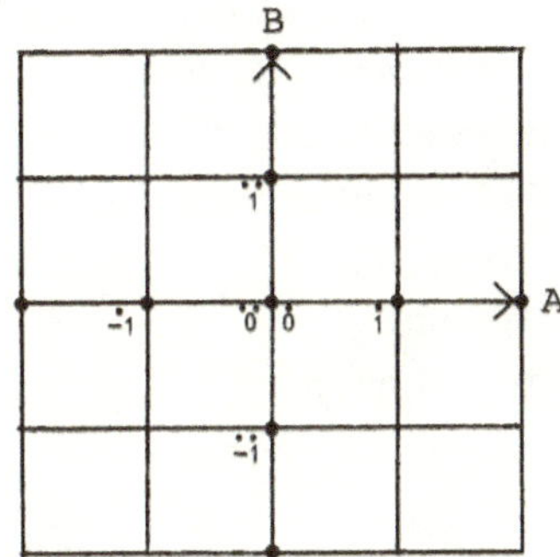

For example, if α = I and β = exp, then *-paper is semi-
log paper that is logarithmically scaled on the vertical axis.

A *-point is an ordered pair of numbers in A and B, in
that order.

Informally, the *-distance, $\overset{*}{d}(P_1,P_2)$, between two *-
points P_1 and P_2 is determined by plotting them on *-paper
and measuring their separation with the 'β-ruler' provided
by the vertical axis. The result is a β-nonnegative number.
(The concept of *-distance will be discussed formally in
Section 10.2.)

The *-graph of a set of *-points is the result of plotting them on *-paper. As expected, the *-graph of every *-
uniform function is a straight line.

Consider the *-graph of a *-uniform function g whose *-
slope is β-positive. Let P_1 and P_2 be two distinct *-points
on g, as shown below, and let P_3 be the vertex of the indicated right triangle.

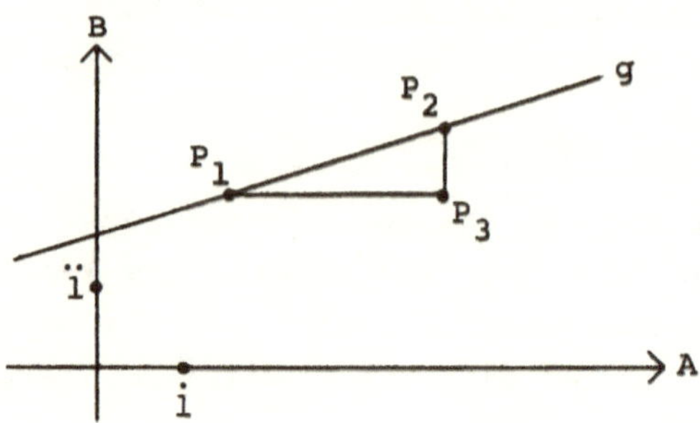

Then the *-slope of g equals $\overset{*}{\mathrm{d}}(P_2,P_3)$ $\overset{..}{/}$ $\overset{*}{\mathrm{d}}(P_1,P_3)$. With obvious adjustments an interpretation may be provided if the *-
slope of g is β-negative. If the *-slope of g is $\overset{..}{0}$, then the
*-graph of g is horizontal.

     In Section 6.3 we observed that the *-derivative of a
function f at an argument a is equal to the *-slope of the *-
uniform function g that is *-tangent to f at (a,f(a)). The
*-graph would look like this:

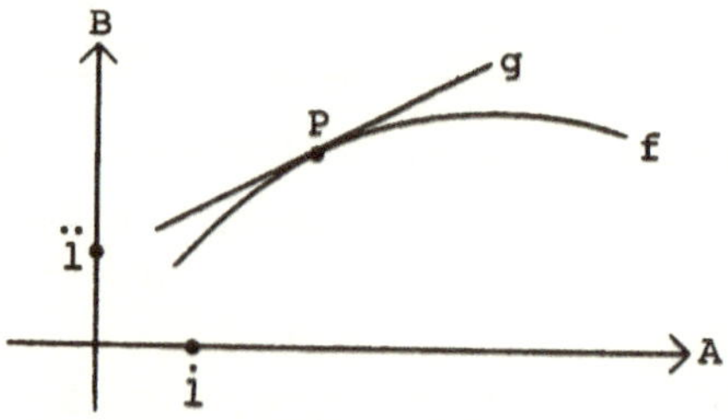

     The purely arithmetic nature of the *-average surely ob-
viates a graphic interpretation thereof.

     Now we interpret the *-integral. All references here to
geometric figures are intended to apply to figures as they
appear on *-paper. The *-area of a rectangle is defined to
be the β-product of its '*-length' and '*-width.' The *-
area of a unit square (i.e., a square whose sides all have
length $\overset{..}{1}$) turns out to be $\overset{..}{1}$. If a rectangle can be decom-
posed into n unit squares, then its *-area equals $\overset{..}{n}$, as expected.

     Let f be a β-positive, *-continuous function on $\overset{.}{[}r,s\overset{.}{]}$,

and let S be the region, on *-paper, bounded by the *-graph
of f, the horizontal axis, and the vertical lines at r and s.
By using α-partitions, rectangles, and the β-limit process,
one can readily define the *-area of S, which turns out to

be $\displaystyle\int_{r}^{\overset{*}{s}} f.$

Another graphic interpretation of the *-calculus may be
obtained by using <u>Cartesian paper</u>, by which we mean paper
that is ruled off in squares and marked thus:

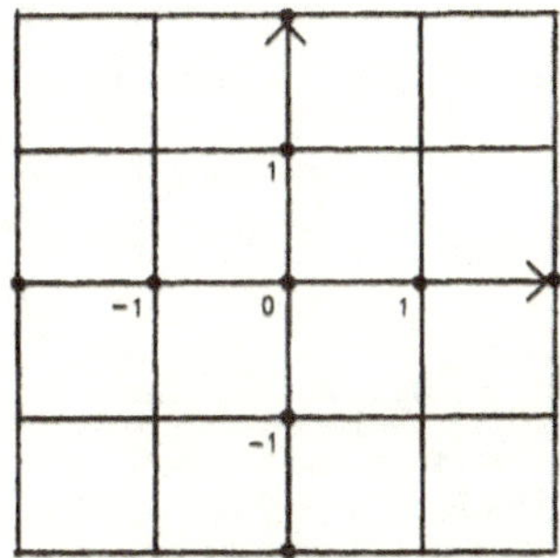

The <u>C-graph</u> of a *-point (a,b) is the graph of the point
$(\alpha^{-1}(a), \beta^{-1}(b))$ on Cartesian paper. The C-graph of a set of
*-points consists of the C-graphs of all the individual *-
points. The C-graph of each *-uniform function is a straight
line.

If P_1 and P_2 are *-points and t is the Euclidean distance between their C-graphs, then the *-distance between P_1
and P_2 equals $\beta(t)$.

If f is a *-uniform function and m is the classical
slope of its C-graph, then the *-slope of f equals $\beta(m)$.

If f is a β-positive, *-continuous function on $[r,s]$,
and w is the (usual) area of the region bounded by the C-
graph of f, the horizontal axis, and the vertical lines at
$\alpha^{-1}(r)$ and $\alpha^{-1}(s)$, then

$$\int_{r}^{\overset{*}{s}} f = \beta(w).$$

Chapter 7

THE QUADRATIC FAMILY OF CALCULI

7.1 THE QUADRATIC ARITHMETIC

In this chapter we obtain three specific calculi from the *-calculus by using the classical and quadratic arithmetics.

<u>Quadratic</u> <u>arithmetic</u> is the arithmetic generated by the function that assigns to each number x the number

$$x^{\overline{\overline{1/2}}} = \begin{cases} \sqrt{x} & \text{if } x > 0 \\ 0 & \text{if } x = 0 \\ -\sqrt{-x} & \text{if } x < 0 \end{cases}$$

That function is on R, onto R, and one-to-one; its inverse assigns to each number x the number

$$x^{\overline{\overline{2}}} = \begin{cases} x^2 & \text{if } x > 0 \\ 0 & \text{if } x = 0 \\ -x^2 & \text{if } x < 0 \end{cases}$$

For any numbers y and z, we have

$$(yz)^{\overline{\overline{1/2}}} = y^{\overline{\overline{1/2}}} \cdot z^{\overline{\overline{1/2}}},$$

$$(y^{\overline{\overline{2}}})^{\overline{\overline{1/2}}} = y = (y^{\overline{\overline{1/2}}})^{\overline{\overline{2}}}.$$

Quadratic arithmetic has the following features. (The numbers y and z are arbitrary.)

```
Realm..................... R
Quadratic zero............ 0
Quadratic one............. 1
```

Quadratic addition........ $y \oplus z = (y^{\overline{\overline{2}}} + z^{\overline{\overline{2}}})^{\overline{\overline{1/2}}}$

Quadratic subtraction..... $y \ominus z = (y^{\overline{\overline{2}}} - z^{\overline{\overline{2}}})^{\overline{\overline{1/2}}}$

```
Quadratic multiplication.. ⎫
Quadratic division........ ⎪  Identical with
Quadratic order...........  ⎬  corresponding
Quadratic intervals.......  ⎪  classical notions
Quadratic convergence..... ⎭
```

Quadratic extent of [y,z]. $z \ominus y$

52

The quadratic average of n numbers $u_1,\ldots,u_n$ turns out to be
$$\left\{(u_1{}^{\bar{\bar{2}}} + \ldots + u_n{}^{\bar{\bar{2}}})/n\right\}^{\overline{1/2}},$$
which reduces to the well-known root mean square when the u_i are non-negative. The definitions of quadratic progressions and quadratic partitions specialize readily from those of α-progressions and α-partitions in Section 5.3.

Quadratic arithmetic is a member of the infinite family of power arithmetics discussed in Note 5.

7.2 THE QUADRATIC CALCULUS

The <u>quadratic calculus</u> is the *-calculus determined by the ordered pair * whose members are classical arithmetic and quadratic arithmetic, in that order.

In the quadratic calculus, the operators are applied to functions with arguments and values in R, *-limits and *-continuity are identical with classical limits and classical continuity, the isomorphism ι is the generator of quadratic arithmetic, and the β-change of a function f over [r,s] is equal to
$$\left\{[f(s)]^{\bar{\bar{2}}} - [f(r)]^{\bar{\bar{2}}}\right\}^{\overline{1/2}},$$
which is negative if $f(s) < f(r)$, zero if $f(s) = f(r)$, and positive if $f(s) > f(r)$.

For the remainder of this section we shall use the adjective "quadratic" instead of the prefix "*."

The quadratically-uniform functions are expressible in the form $(mx + c)^{\overline{1/2}}$, where m and c are constants and x is unrestricted in R. The quadratic slope of the preceding function equals $m^{\overline{1/2}}$.

The quadratic gradient of f over [r,s] equals
$$\left\{\frac{[f(s)]^{\bar{\bar{2}}} - [f(r)]^{\bar{\bar{2}}}}{s - r}\right\}^{\overline{1/2}}.$$

The quadratic derivative of f at a, denoted by $[\overset{\circ}{D}f](a)$, coexists with $[D(f^{\bar{\bar{2}}})](a)$, and

            The Quadratic Family of Calculi

$$[\overset{\circ}{D}f](a) = \left\{[D(f^{\overline{\overline{2}}})](a)\right\}^{\overline{\overline{1/2}}}.$$

If $[Df](a)$ exists, then $[\overset{\circ}{D}f](a)$ exists; and if $[\overset{\circ}{D}f](a)$ exists and $f(a) \neq 0$, then $[Df](a)$ exists. Moreover, $[\overset{\circ}{D}f](a)$ exists if and only if f has a quadratic tangent T at $(a,f(a))$. If $[\overset{\circ}{D}f](a)$ does exist, it equals the quadratic slope of T, and if, furthermore, $f(a) \neq 0$, then T is classically tangent to f at $(a,f(a))$.

Of course, $\overset{\circ}{D}f$ is constant on R if and only if f is quadratically-uniform.

The quadratic average of a continuous function f on $[r,s]$ equals

$$\left\{\frac{1}{s-r} \int_r^s (f^{\overline{\overline{2}}})\right\}^{\overline{\overline{1/2}}},$$

which reduces to the root mean square when f is non-negative.

The quadratic integral of a continuous function f on $[r,s]$, denoted by $\overset{\circ}{\int_r^s} f$, equals

$$\left\{\int_r^s (f^{\overline{\overline{2}}})\right\}^{\overline{\overline{1/2}}},$$

which, for f non-negative, reduces to

$$\left\{\int_r^s (f^2)\right\}^{1/2}.$$

That expression appears often in the mathematical literature, though we have never seen it identified as an integral.

All the operators of the quadratic calculus are homogenous and quadratically additive; for example,

$$\overset{\circ}{D}(c \cdot f) = c \cdot \overset{\circ}{D}f, \quad c \text{ constant,}$$

$$\overset{\circ}{D}(f \oplus g) = \overset{\circ}{D}f \oplus \overset{\circ}{D}g.$$

The Basic Theorem of Quadratic Calculus

If $\overset{\circ}{D}f$ is continuous on $[r,s]$, then its quadratic average on $[r,s]$ equals the quadratic gradient of f over $[r,s]$.

The First Fundamental Theorem of Quadratic Calculus

If f is continuous on [r,s], and

$$g(x) = \overset{o}{\int}_{r}^{x} f, \quad \text{for every number x in [r,s],}$$

then

$$\overset{o}{D}g = f, \quad \text{on [r,s].}$$

The Second Fundamental Theorem of Quadratic Calculus

If $\overset{o}{D}h$ is continuous on [r,s], then

$$\overset{o}{\int}_{r}^{s} (\overset{o}{D}h) = h(s) \ominus h(r).$$

7.3 THE ANAQUADRATIC CALCULUS

The anaquadratic calculus is the *-calculus determined
by the ordered pair * whose members are quadratic arithmetic
and classical arithmetic, in that order.

In the anaquadratic calculus the operators are applied
to functions with arguments and values in R, *-limits and *-
continuity are identical with classical limits and classical
continuity, the isomorphism ι is the inverse of the generator of quadratic arithmetic, and the β-change of a function
f over [r,s] equals f(s) - f(r), since here $\beta = I$.

For the remainder of this section we shall use the adjective "anaquadratic" instead of the prefix "*."

The anaquadratically-uniform functions are expressible
in the form $mx^{\overline{\overline{2}}} + c$, where m and c are constants and x is
unrestricted in R. The anaquadratic slope of the preceding
function equals m.

The anaquadratic gradient of f over [r,s] equals

$$\frac{f(s) - f(r)}{s^{\overline{\overline{2}}} - r^{\overline{\overline{2}}}} .$$

The anaquadratic derivative of f at a, denoted by
[Df](a), coexists with, and is equal to, the classical de-
rivative of the function $\overline{\overline{f}}(x) = f(x^{\overline{1/2}})$ at $a^{\overline{\overline{2}}}$:

$$[\overset{}{D}f](a) = [D\overline{\overline{f}}](a^{\overline{\overline{2}}}).$$

If $[\underset{o}{D}f](a)$ exists, then $[Df](a)$ exists; and if $[Df](a)$ exists and $a \neq 0$, then $[\underset{o}{D}f](a)$ exists. Moreover, $[\underset{o}{D}f](a)$ exists if and only if f has an anaquadratic tangent T at $(a, f(a))$. If $[\underset{o}{D}f](a)$ does exist, it equals the anaquadratic slope of T, and if, furthermore, $a \neq 0$, then T is classically tangent to f at $(a, f(a))$.

Also worth noting is the fact that $[\underset{o}{D}f](a)$, if it exists, equals the classical derivative, at a, of f with respect to the function $h(x) = x^2$.

Of course, $\underset{o}{D}f$ is constant on R if and only if f is anaquadratically-uniform.

Let f be continuous on $[r,s]$, and set $\bar{f}(x) = f(x^{1/2})$, $\bar{r} = r^2$, and $\bar{s} = s^2$. Then the anaquadratic average of f on $[r,s]$ equals

$$\frac{1}{\bar{s} - \bar{r}} \int_{\bar{r}}^{\bar{s}} \bar{f},$$

and the anaquadratic integral of f on $[r,s]$, denoted by $\underset{o}{\int_{r}^{s}} f$,

equals $\int_{\bar{r}}^{\bar{s}} \bar{f}$, which is the Stieltjes integral of f with respect to the function $h(x) = x^2$ on $[r,s]$.

All the operators of the anaquadratic calculus are additive and homogeneous.

The Basic Theorem of Anaquadratic Calculus

If $\underset{o}{D}f$ is continuous on $[r,s]$, then its anaquadratic average on $[r,s]$ equals the anaquadratic gradient of f over $[r,s]$.

The First Fundamental Theorem of Anaquadratic Calculus

If f is continuous on $[r,s]$, and

$$g(x) = \underset{o}{\int_{r}^{x}} f, \quad \text{for every number x in } [r,s],$$

then

$$\underset{o}{D}g = f, \quad \text{on } [r,s].$$

The Second Fundamental Theorem of Anaquadratic Calculus

If $\overset{\circ}{D}h$ is continuous on $[r,s]$, then

$$\int_{\overset{}{\underset{\circ}{r}}}^{s} (\overset{\circ}{D}h) = h(s) - h(r).$$

7.4 THE BIQUADRATIC CALCULUS

The biquadratic calculus is the $*$-calculus determined by the ordered pair $*$ both of whose members are quadratic arithmetic.

In the biquadratic calculus the operators are applied to functions with arguments and values in R, $*$-limits and $*$-continuity are identical with classical limits and classical continuity, the isomorphism ι is the identity function I, and the β-change of a function f over $[r,s]$ equals

$$\left\{ [f(s)]^{\overline{\overline{2}}} - [f(r)]^{\overline{\overline{2}}} \right\}^{\overline{1/2}}.$$

For the remainder of this section we shall use the adjective "biquadratic" instead of the prefix "$*$."

The biquadratically-uniform functions are expressible in the form $(mx^{\overline{\overline{2}}} + c)^{\overline{1/2}}$, where m and c are constants and x is unrestricted in R. The biquadratic slope of the preceding function equals $m^{\overline{1/2}}$.

The biquadratic gradient of f over $[r,s]$ equals

$$\left\{ \frac{[f(s)]^{\overline{\overline{2}}} - [f(r)]^{\overline{\overline{2}}}}{s^{\overline{\overline{2}}} - r^{\overline{\overline{2}}}} \right\}^{\overline{1/2}}.$$

The biquadratic derivative of f at a, denoted by $[\overset{\circ}{D}f](a)$, coexists with $[D\overline{\overline{f}}](\overline{a})$ where $\overline{\overline{f}}(x) = [f(x^{\overline{1/2}})]^{\overline{\overline{2}}}$ and $\overline{a} = a^{\overline{\overline{2}}}$; and

$$[\overset{\circ}{D}f](a) = \{ [D\overline{\overline{f}}](\overline{a}) \}^{\overline{1/2}}.$$

If $a \neq 0$ and $f(a) \neq 0$, then $[Df](a)$ and $[\overset{\circ}{D}f](a)$ coexist. Moreover, $[\overset{\circ}{D}f](a)$ exists if and only if f has a biquadratic tangent T at $(a,f(a))$. If $[\overset{\circ}{D}f](a)$ does exists, it equals the biquadratic slope of T, and if, furthermore, $a \neq 0$ and $f(a) \neq$

0, then T is classically tangent to f at (a,f(a)).

Of course, $\overset{\circ}{D}f$ is constant on R if and only if f is bi-quadratically-uniform.

Let f be continuous on [r,s], and set

$$\bar{f}(x) = [f(x^{\overline{1/2}})]^{\overline{2}},$$

$\bar{r} = r^{\overline{2}}$, and $\bar{s} = s^{\overline{2}}$. Then the biquadratic average of f on [r,s] equals

$$\left\{ \frac{1}{\bar{s} - \bar{r}} \int_{\bar{r}}^{\bar{s}} \bar{f} \right\}^{\overline{1/2}},$$

and the biquadratic integral of f on [r,s], denoted by $\overset{\circ}{\int}_{r}^{s} f$, equals

$$\left\{ \int_{\bar{r}}^{\bar{s}} \bar{f} \right\}^{\overline{1/2}}.$$

All the operators of the biquadratic calculus are homogeneous and quadratically additive.

The Basic Theorem of Biquadratic Calculus

If $\overset{\circ}{D}f$ is continuous on [r,s], then its biquadratic average on [r,s] equals the biquadratic gradient of f over [r,s].

The First Fundamental Theorem of Biquadratic Calculus

If f is continuous on [r,s], and

$$g(x) = \overset{\circ}{\int_{r}^{x}} f, \quad \text{for every number x in [r,s],}$$

then

$$\overset{\circ}{D}g = f, \quad \text{on [r,s].}$$

The Second Fundamental Theorem of Biquadratic Calculus

If $\overset{\circ}{D}h$ is continuous on [r,s], then

$$\overset{\circ}{\int_{r}^{s}} (\overset{\circ}{D}h) = h(s) \ominus h(r).$$

Chapter 8
THE HARMONIC FAMILY OF CALCULI

8.1 THE HARMONIC ARITHMETIC

In this chapter we obtain three calculi from the $*$-calculus by using the classical and harmonic arithmetics.

Harmonic arithmetic is the arithmetic generated by the function that assigns to each number x the number

$$\frac{1}{x} = \begin{cases} 1/x & \text{if} \quad x \neq 0 \\ 0 & \text{if} \quad x = 0 \end{cases}.$$

We shall also use the notation $1/\!/x$.

The preceding function is on R, onto R, one-to-one, and identical with its inverse, since

$$1/\!/(1/\!/x) \;=\; x,$$

for every number x.

Harmonic arithmetic has the following features. (The numbers y and z are arbitrary.)

Realm....................... R

Harmonic zero.............. 0

Harmonic one.............. 1

Harmonic addition......... y $\boxed{+}$ z = $1/\!/(1/\!/y + 1/\!/z)$
Harmonic subtraction...... y $\boxed{-}$ z = $1/\!/(1/\!/y - 1/\!/z)$
Harmonic multiplication... ⎰Identical with corresponding
Harmonic division......... ⎱classical notions.

Harmonic order............ y $\boxed{<}$ z if and only if $1/\!/y < 1/\!/z$.

Harmonic interval, $\boxed{[}y,z\boxed{]}$.. All numbers x such that y $\boxed{\leq}$ x $\boxed{\leq}$ z.

Harmonic extent of $\boxed{[}y,z\boxed{]}$.. z $\boxed{-}$ y

The following facts should also be noted: the symbol $/\!/$ does not stand for harmonic division, which is identical with classical division; harmonic order is not identical with classical order (for example, -1 $\boxed{<}$ -2); the generator of harmonic arithmetic is not classically continuous at 0; and harmonic convergence is not identical with classical convergence (for instance, the infinite sequence {n} is harmonically convergent to 0).

59

Harmonic sums occur frequently in science. For example, if two resistors of resistance y and z are connected in parallel, the combined resistance equals the harmonic sum of y and z.

Since the natural average and natural progressions in harmonic arithmetic are direct extensions of the well-known harmonic average and harmonic progressions defined for positive numbers, we shall use the same names for the extensions.

The harmonic average of n numbers $u_1,\ldots,u_n$ equals

$$\frac{1}{\dfrac{(1/\!/u_1) + \ldots + (1/\!/u_n)}{n}} \, .$$

Harmonic arithmetic is a member of the infinite family of power arithmetics discussed in Note 5.

8.2 THE HARMONIC CALCULUS

The <u>harmonic calculus</u> is the *-calculus determined by the ordered pair * whose members are classical arithmetic and harmonic arithmetic, in that order.

In the harmonic calculus the operators are applied to functions with arguments and values in R, *-limits and *-continuity are NOT always identical with classical limits and classical continuity, the isomorphism ι is the generator of harmonic arithmetic, and the β-change of a function f over [r,s] equals

$$\frac{1}{1/\!/f(s) - 1/\!/f(r)},$$

which is 0 when f(s) = f(r).

For the remainder of this section we shall use the adjective "harmonic" instead of the prefix "*."

The harmonically-uniform functions are expressible in the form $1/\!/(mx + c)$, where m and c are constants and x is unrestricted in R. The harmonic slope of the preceding function equals $1/\!/m$.

The harmonic gradient of f over [r,s] equals

$$\frac{1}{\dfrac{1/\!\!/f(s) \; - \; 1/\!\!/f(r)}{s \; - \; r}} \; .$$

The harmonic derivative of f at a, denoted by $[\overset{\square}{D}f](a)$, coexists with $[D(1/\!\!/f)](a)$, and

$$[\overset{\square}{D}f](a) \;\; = \;\; \frac{1}{[D(1/\!\!/f)](a)} \; .$$

If $f(a) \neq 0$, then $[Df](a)$ and $[\overset{\square}{D}f](a)$ coexist. Moreover, $[\overset{\square}{D}f](a)$ exists if and only if f has a harmonic tangent T at $(a,f(a))$. If $[\overset{\square}{D}f](a)$ does exist, it equals the harmonic slope of T, and if, furthermore, $f(a) \neq 0$, then T is classically tangent to f at $(a,f(a))$.

Of course, $\overset{\square}{D}f$ is constant on R if and only if f is harmonically-uniform.

The harmonic average of a harmonically continuous function f on $[r,s]$ equals

$$\frac{1}{\dfrac{1}{s \; - \; r} \displaystyle\int_{r}^{s} (1/\!\!/f)} \; ,$$

which reduces to the traditional harmonic average when f is positive.

The harmonic integral of a harmonically continuous function f on $[r,s]$, denoted by $\displaystyle\overset{\square}{\int_{r}^{s}} f$, equals

$$\frac{1}{\displaystyle\int_{r}^{s} (1/\!\!/f)} \; .$$

All the operators of the harmonic calculus are homogeneous and harmonically additive; for example,

$$D(c\cdot f) \; = \; c\cdot Df, \quad c \text{ constant,}$$

$$D(f \boxplus g) \; = \; Df \boxplus Dg.$$

The Basic Theorem of Harmonic Calculus

If $\overset{\square}{D}f$ is harmonically continuous on $[r,s]$, then
its harmonic average on $[r,s]$ equals the harmonic
gradient of f over $[r,s]$.

The First Fundamental Theorem of Harmonic Calculus

If f is harmonically continuous on $[r,s]$, and

$$g(x) = \int_{r}^{\overset{\square}{}x} f, \quad \text{for every number x in } [r,s],$$

then

$$\overset{\square}{D}g = f, \quad \text{on } [r,s].$$

The Second Fundamental Theorem of Harmonic Calculus

If $\overset{\square}{D}h$ is harmonically continuous on $[r,s]$, then

$$\int_{r}^{\overset{\square}{}s} (\overset{\square}{D}h) = h(s) \boxminus h(r).$$

8.3 THE ANAHARMONIC CALCULUS

The anaharmonic calculus is the *-calculus determined
by the ordered pair * whose members are harmonic arithmetic
and classical arithmetic, in that order.

In the anaharmonic calculus the operators are applied
to functions with arguments and values in R, *-limits and
*-continuity are NOT always identical with classical limits
and classical continuity, the isomorphism ι is the (self-
inverse) generator of harmonic arithmetic, and the β-change
of a function f over $[\overset{\square}{r},\overset{\square}{s}]$ equals $f(s) - f(r)$, since here
β = I.

For the remainder of this section we shall use the adjective "anaharmonic" instead of the prefix "*."

The anaharmonically-uniform functions are expressible in
the form $m(1/\!/x) + c$, where m and c are constants and x is unrestricted in R. The anaharmonic slope of the preceding function equals m.

The anaharmonic gradient of f over $[\overset{\square}{r},\overset{\square}{s}]$ equals

$$\frac{f(s) - f(r)}{1/\!/s - 1/\!/r}.$$

The anaharmonic derivative of f at a, denoted by $[\underset{\square}{D}f](a)$, coexists with, and is equal to, the classical derivative of the function $\bar{f}(x) = f(1/\!/x)$ at $1/\!/a$:

$$[\underset{\square}{D}f](a) = [D\bar{f}](1/\!/a).$$

If $a \neq 0$, then $[Df](a)$ and $[\underset{\square}{D}f](a)$ coexist. Moreover, $[\underset{\square}{D}f](a)$ exists if and only if f has an anaharmonic tangent T at $(a,f(a))$. If $[\underset{\square}{D}f](a)$ does exist, it equals the anaharmonic slope of T, and if, furthermore, $a \neq 0$, then T is classically tangent to f at $(a,f(a))$.

Of course, $\underset{\square}{D}f$ is constant on R if and only if f is anaharmonically-uniform.

Let f be anaharmonically continuous on $[r,s]^{\square}_{\square}$, and set $\bar{f}(x) = f(1/\!/x)$, $\bar{r} = 1/\!/r$, and $\bar{s} = 1/\!/s$. Then the anaharmonic average of f on $[r,s]^{\square}_{\square}$ equals

$$\frac{1}{\bar{s} - \bar{r}} \int_{\bar{r}}^{\bar{s}} \bar{f},$$

and the anaharmonic integral of f on $[r,s]^{\square}_{\square}$, denoted by $\int_{\square r}^{s} f$, equals $\int_{\bar{r}}^{\bar{s}} \bar{f}$.

All the operators of the anaharmonic calculus are additive and homogeneous.

The Basic Theorem of Anaharmonic Calculus

If $\underset{\square}{D}f$ is anaharmonically continuous on $[r,s]^{\square}_{\square}$, then its anaharmonic average on $[r,s]^{\square}_{\square}$ equals the anaharmonic gradient of f over $[r,s]^{\square}_{\square}$.

The First Fundamental Theorem of Anaharmonic Calculus

If f is anaharmonically continuous on $[r,s]^{\square}_{\square}$, and

$$g(x) = \int_{\square r}^{x} f, \quad \text{for every number x in } [r,s]^{\square}_{\square},$$

then

$$\underset{\square}{D}g = f, \quad \text{on } [r,s]^{\square}_{\square}.$$

The Second Fundamental Theorem of Anaharmonic Calculus

If $\underset{\square}{D}h$ is anaharmonically continuous on $[r,s]^{\square}_{\square}$, then

$$\int_{\square r}^{s} (\underset{\square}{D}h) = h(s) - h(r).$$

8.4 THE BIHARMONIC CALCULUS

The biharmonic calculus is the *-calculus determined by
the ordered pair * both of whose members are harmonic arithmetic.

In the biharmonic calculus the operators are applied to
functions with arguments and values in R, *-limits and *-con-
tinuity are NOT always identical with classical limits and
classical continuity, the isomorphism ι is the identity function I, and the β-change of a function f over $[r,s]$ equals

$$\frac{1}{1/\!/f(s) \ - \ 1/\!/f(r)}.$$

For the remainder of this section we shall use the adjective "biharmonic" instead of the prefix "*."

The biharmonically-uniform functions are expressible in
the form $1/\!/[m(1/\!/x) + c]$, where m and c are constants and x
is unrestricted in R. The biharmonic slope of the preceding
function equals $1/\!/m$.

The biharmonic gradient of f over $[r,s]$ equals

$$\frac{\dfrac{1}{1/\!/f(s) \ - \ 1/\!/f(r)}}{1/\!/s \ - \ 1/\!/r}$$

The biharmonic derivative of f at a, denoted by $[Df](a)$,
coexists with $[D\bar{f}](\bar{a})$, where $\bar{f}(x) = 1/\!/f(1/\!/x)$ and $\bar{a} = 1/\!/a$; and

$$[Df](a) \ = \ 1/\!/[D\bar{f}](\bar{a}).$$

If $a \neq 0$ and $f(a) \neq 0$, then $[Df](a)$ and $[Df](a)$ coexist.
Moreover, $[Df](a)$ exists if and only if f has a biharmonic
tangent T at $(a,f(a))$. If $[Df](a)$ does exist, it equals the
biharmonic slope of T, and if, furthermore, $a \neq 0$ and $f(a) \neq$
0, then T is classically tangent to f at $(a,f(a))$.

Of course, Df is constant on R if and only if f is biharmonically-uniform.

Let f be biharmonically continuous on $[r,s]$, and set
$\bar{f}(x) = 1/\!/f(1/\!/x)$, $\bar{r} = 1/\!/r$, and $\bar{s} = 1/\!/s$. Then the biharmonic

average of f on $[r,s]$ equals

$$\frac{1}{\dfrac{1}{\bar{s}-\bar{r}}\displaystyle\int_{\bar{r}}^{\bar{s}}\bar{f}},$$

and the biharmonic integral of f on $[r,s]$, denoted by $\displaystyle\int_r^s f$, equals

$$\frac{1}{\displaystyle\int_{\bar{r}}^{\bar{s}}\bar{f}}.$$

All the operators of the biharmonic calculus are homogeneous and harmonically additive.

The Basic Theorem of Biharmonic Calculus

If Df is biharmonically continuous on $[r,s]$, then its biharmonic average on $[r,s]$ equals the biharmonic gradient of f over $[r,s]$.

The First Fundamental Theorem of Biharmonic Calculus

If f is biharmonically continuous on $[r,s]$, and

$$g(x) = \int_r^x f, \quad \text{for every number } x \text{ in } [r,s],$$

then

$$Dg = f, \quad \text{on } [r,s].$$

The Second Fundamental Theorem of Biharmonic Calculus

If Dh is biharmonically continuous on $[r,s]$, then

$$\int_r^s (Dh) = h(s) \boxminus h(r).$$

Chapter 9
H E U R I S T I C S

<u>9.1 INTRODUCTION</u>

Since it is not unreasonable to suppose that there are situations where a non-Newtonian calculus might be useful, we shall propose some guides that may be helpful in selecting appropriate gradients, derivatives, integrals, and averages. Of course, one is always free to use any operator that is meaningful in a given context.

We continue to let $*$ be an ordered pair of arbitrarily chosen arithmetics, $(A,\dot{+},\dot{-},\dot{\times},\dot{/},\dot{<})$ and $(B,\ddot{+},\ddot{-},\ddot{\times},\ddot{/},\ddot{<})$, with generators α and β, respectively.

<u>9.2 CHOOSING GRADIENTS AND DERIVATIVES</u>

Since the choice of a derivative tacitly involves a choice of the related gradient, we shall restrict our attention to gradients, for which three heuristic principles will be offered.

<u>Principle I</u>

> If the natural methods of measuring changes in arguments and values are provided by α-differences and β-differences, respectively, then the $*$-gradient may be appropriate.

For example, the classical gradient is appropriate for analyzing motion because changes in time and position are naturally measured by differences. The geometric gradient is appropriate for analyzing stock-price movements because changes in time and price are usually measured by differences and geometric differences (i.e., ratios). The bigeometric gradient may be useful in psychophysics because changes in stimulus and sensation are often measured by geometric differences. ("Equal stimulus ratios produce equal sensation ratios," according to S.S. Stevens.)

<u>Principle II</u>

> If the functional relationship between two magnitudes would be, or is assumed to be, $*$-uniform under normal

or ideal conditions, then the *-gradient may be appropriate.

For example, the geometric gradient may be appropriate for analyzing radioactive decay because the relationship of mass to time is assumed to be geometrically-uniform (i.e., exponential) under ideal conditions. Similarly, the sigmoidal gradient may be useful in the study of growth that is normally sigmoidally-uniform (Note 4). The harmonic gradient may be appropriate for analyzing the relationship between two magnitudes that would normally be inversely proportional. Finally we observe that another reason for using the classical gradient in the study of motion is the assumption that under ideal conditions, i.e., the absence of forces, the relationship of position to time would be linear (Newton's First Law).

It is worth noting that each one-to-one function f on R is *-uniform if one chooses $\alpha = I$ and $\beta = f$. This implies that each such function has a derivative that is constant on R.

The statement of the third principle requires some preliminary definitions.

An α-translation is a function of the form $x \overset{\bullet}{+} k$, where k is a constant in A and x is unrestricted in A. For example, the classical translations have the form $x + k$, and the quadratic translations have the form

$$\{x^{\overline{\overline{2}}} + k^{\overline{\overline{2}}}\}^{\overline{\overline{1/2}}}.$$

Each α-translation $x \overset{\bullet}{+} k$ transforms $\overset{\bullet}{0}$ to k and preserves α-differences, that is, for any numbers u and v in A, the α-difference between u and v equals the α-difference between the transforms of u and v. Therefore one may say that each α-translation effects a shift in origin.

An α-similitude is a function of the form $p \overset{\bullet}{\times} x$, where p is an α-positive constant and x is an arbitrary α-positive number. Since each α-similitude $p \overset{\bullet}{\times} x$ transforms $\overset{\bullet}{1}$ to p and preserves α-quotients, one may say that α-similitudes effect a change in unit.

Of course, β-translations and β-similitudes are defined similarly.

The classical similitudes are exp-translations, that is, geometric translations. Indeed, every similitude is a translation (each α-similitude is a γ-translation, where $\gamma(x) = \alpha(e^x)$).

It is an important fact that the *-gradient (and *-derivative) is invariant under all α-translations of the arguments and all β-translations of the values. For example, the classical gradient is invariant under all classical translations of arguments and of values, and the geometric gradient is invariant under all classical translations of arguments and all geometric translations of values.

<u>Principle III</u>

> If one desires a gradient that is invariant under
> all α-translations of arguments and all β-transla-
> tions of values, then one should consider using
> the *-gradient.

It is an interesting fact that the mixed second derivative $\widetilde{D}(D)$ is invariant under all classical translations and classical similitudes of values, and $\underset{\sim}{\widetilde{D}}(D)$ is invariant under all classical translations and classical similitudes of arguments and of values.

<u>9.3 CHOOSING INTEGRALS</u>

Suppose that a scientific proposition has been expressed by a *-differential equation, that is, a differential equation involving *-derivatives and perhaps related arithmetic operations. To solve the equation one would certainly try to use *-integration. If obtained, the solutions would at first be expressed by means of *-integrals with variable upper limits, which in turn would be evaluated in terms of known functions or new functions defined for the purpose. (Many functions were originally defined as classical integrals.) It is even conceivable that solutions in closed form could be obtained for certain intractable classical differential equations by re-expressing them with appropriate non-Newtonian derivatives and related arithmetic operations.

A differential equation containing several types of derivatives can readily be transformed into an equation involving only classical derivatives or only non-Newtonian derivatives of the same type. (See Sections 6.8 and 6.9.)

Integrals are also used for defining scientific concepts. Consider, for example, the physicists' concept of work. If the force f is constant, the work on the position interval [r,s] is defined to be (s - r)·f, which we denote here by $W_r^s f$. When f is constant the following conditions are clearly satisfied.

1. $W_r^s f = (s - r) \cdot f.$

2. Work is monotonic increasing with respect to force.

3. Work is additive with respect to displacement; that is, for any positions r,s,t such that r < s < t,

$$W_r^s f + W_s^t f = W_r^t f.$$

Desiring to extend the concept of work to the case where the force f is continuously variable, the physicist stipulates that the extended work concept should satisfy Condition 1 when f is constant and should satisfy Conditions 2 and 3 in general. The solution is now uniquely determined; for according to Section 1.6, the operator W satisfies those three conditions if and only if W is the classical integral. Thus, the physicist must adopt the following definition:

$$W_r^s f = \int_r^s f.$$

The foregoing well-known example illustrates the fact that the characterization of the classical integral in Section 1.6 is a heuristic guide for its appropriate use.

Now suppose that a scientist is concerned with a positive magnitude g, called 'gorce,' which may be continuously variable with respect to time. If g is constant, he defines the 'toil' on the time interval [r,s] to be g^{s-r}, denoted by $T_r^s g$. (Somewhere we have seen such a concept, perhaps in economics.) When g is constant, the following conditions are clearly satisfied.

1. $T_r^s g = g^{s-r}$.

2. Toil is monotonic increasing with respect to gorce.

3. Toil is multiplicative with respect to time; that
   is, for any instants r,s,t such that r < s < t,

$$T_r^s g \cdot T_s^t g = T_r^t g.$$

To extend the toil concept to the case where the gorce
g is continuously variable, we stipulate that the extended
toil concept should satisfy Condition 1 when g is constant
and should satisfy Conditions 2 and 3 in general. The solution is now uniquely determined; for, according to Section
2.6, the operator T satisfies those three conditions if and
only if T is the geometric integral. Thus, we must adopt
the following definition:

$$T_r^s g = \int_r^{\tilde{s}} g.$$

It should now be clear that the characterization of the
*-integral in Section 6.6 is a heuristic guide for the use
of integrals.

9.4 CHOOSING AVERAGES

Most of this section is devoted to β-averages rather
than *-averages, since the choice of the latter depends primarily on how one would average finite sequences of values,
although the method of partitioning the argument intervals
is certainly a factor. (In contrast to the *-integral, the
*-average is not independent of the type of partitions used
for the domain.)

First let it be noted that our definition of the β-
average corresponds to a definition of the arithmetic average of $u_1,\ldots,u_n$ as the number u such that

$$\underbrace{u + \ldots + u}_{n \text{ terms}} = u_1 + \ldots + u_n.$$

Accordingly, the β-average has the same properties relative
to β-arithmetic as the arithmetic average has relative to
classical arithmetic. (One important illustration of that

fact is given near the end of Section 10.4.)

The predominant use of the arithmetic average may be attributed, perhaps, to the custom of measuring deviations by differences rather than by ratios or some other β-difference.   However, any statistical justification for the use of the arithmetic average relative to classical arithmetic can be matched by a justification for the use of the β-average relative to β-arithmetic.

Scientists often invoke a kind of "normalcy" principle in choosing averages. For instance, it is commonly stated that the geometric average is appropriate in the study of those populations which normally increase in geometric progression. This idea can be extended as follows: If under normal or ideal conditions the measurements would be in β-progression, then the β-average may be appropriate.

Some scientists, notably the psychophysicist S.S.Stevens, favor the use of invariance principles for choosing averages. An interesting discussion may be found in the book <u>Basic Concepts of Measurement</u> by Brian Ellis.

Now consider n measurements $u_1,\ldots,u_n$ of magnitudes that naturally combine by β-sums; that is, the measure of a combination of two or more magnitudes equals the β-sum of their individual measures. Surely, then, the β-average of the u_i would be physically meaningful, though the arithmetic average of the u_i might well be physically meaningless. For instance, since the total resistance of n resistors connected in parallel equals the harmonic sum of the individual resistances, one would expect their harmonic average to be physically meaningful.

The intended use of an average is, of course, a basic factor in its choice. Suppose, for example, that $u_1,\ldots,u_n$ are the measurements of one side of a square whose area is to be estimated. The standard procedure is to square the arithmetic average of the u_i. But a different estimate of the area is usually obtained if one computes the arithmetic average of the squares of the u_i. Although some consider that to be no serious objection, it should be noted that the geometric average always yields identical answers both ways.

The point here is that since the area is the square of a side, one should seriously consider using a multiplicative average rather than an additive average.

As is the case for integrals, a fundamental heuristic guide for choosing averages is provided by the three characterizing properties stated at the end of Section 6.4.

In many situations, averages are intuitively more satisfying than integrals. Consider, for instance, a particle moving rectilinearly with positive velocity v at time t. The distance s traveled in the time interval [a,b] is given by

$$s = \int_a^b v.$$

Although the derivation of this result may be clear to a student, he may nevertheless find that the following formula conveys a more immediate meaning:

$$s = (b - a) \cdot M_a^b v;$$

that is, the distance traveled equals the product of the time elapsed and the arithmetic average of the velocity. This version is a direct extension of the case where v is constant. Other examples will readily occur to the reader. (For the last example of the preceding section, we may write

$$T_r^s g = [\tilde{M}_r^s g]^{s-r},$$

a formula that is a direct extension of the case where g is constant.)

9.5 CONSTANTS AND SCIENTIFIC CONCEPTS

In this section we illustrate the thesis that the invention of a scientific concept may depend on the isolation or discovery of a suitable constant. We also suggest that new scientific concepts may arise from the constants provided by the slopes of uniform functions.

Consider the concept of average speed. The definition "distance traveled per unit time" is incomplete because it fails to provide a method of determining the average speed

of an accelerated particle. The definition "distance divided
by time," though not incorrect, is a gross oversimplification
that fails to reveal the underlying issues. Fortunately
there is a completely satisfactory definition, which was un-
doubtedly known to Galileo.

Let us begin with Galileo's definition of uniform motion
as "one in which the distances traversed...during ANY equal
intervals of time are themselves equal." (Continuity of the
motion is tacitly intended here.) Then we isolate a constant
in each given uniform motion by defining speed to be the distance traveled in any unit time interval. Finally, for a
particle that moves non-uniformly a distance d in time t, we
define the average speed to be the speed that a particle in
uniform motion must have in order to travel a distance d in
time t. In our opinion, neither the simplicity nor the obvi-
ousness of the answer, d/t, justifies its use as the definition of average speed.

The critical first step in defining average speed is the
isolation of the constant (speed) in the phenomenon of uniform
motion. Similarly, the critical step in defining the *-gra-
dient is the identification of the constant (*-slope) in a
*-uniform function.

Wherever a scientific phenomenon (or idealized version
thereof) is describable by a *-uniform function, one automat-
ically has a fundamental constant, the *-slope, which may
prove to be useful.

For example, in the idealized model of radioactive decay,
the relationship of mass to time is describable by a geometrically-uniform function whose geometric slope provides a
constant that is independent of the unit of mass and equals
1 more than the percent change of the mass over any unit time
interval.

In the idealized model of the behavior of a gas, the relationship of pressure to volume, assuming temperature constant, is describable by a harmonically-uniform function
whose harmonic slope is a constant that equals the harmonic
change in pressure corresponding to a unit change in volume.

As a final example, consider a fixed light source radiating light uniformly in all directions. Since the intensity of illumination is inversely proportional to the square of the distance from the source, the relationship of intensity to distance is describable by a bigeometrically-uniform function, whose bigeometric slope is independent of the units of distance and intensity.

Chapter 10

COLLATERAL ISSUES

10.1 INTRODUCTION

This last chapter contains, among other things, brief discussions of some ideas that we have entertained but not investigated in depth.

As before, $*$ is an ordered pair of arbitrarily selected arithmetics, $(A, \dot{+}, \dot{-}, \dot{\times}, \dot{/}, \dot{<})$ and $(B, \ddot{+}, \ddot{-}, \ddot{\times}, \ddot{/}, \ddot{<})$, having generators α and β. The isomorphism from α-arithmetic to β-arithmetic is denoted by ι, which was discussed in Section 6.1.

The α-absolute value of a number a in A is denoted by $\dot{|}a\dot{|}$ and equals $\alpha(|\alpha^{-1}(a)|)$; similarly, the β-absolute value of a number b in B is denoted by $\ddot{|}b\ddot{|}$ and equals $\beta(|\beta^{-1}(b)|)$, (Section 5.3).

The $\underline{\beta\text{-square}}$ of a number b in B is $b\ddot{\times}b$, which will be denoted by $b^{\ddot{2}}$ at a slight risk that it might be incorrectly interpreted as b to the power $\beta(2)$. For each β-nonnegative number t, the symbol $\ddot{\sqrt{}}t$ will be used to denote

$$\beta\{\sqrt{\beta^{-1}(t)}\},$$

which is the unique β-nonnegative number whose β-square equals t. For each number b in B,

$$\ddot{\sqrt{b^{\ddot{2}}}} = \ddot{|}b\ddot{|}.$$

For each integer n, the symbols $\dot{n}$ and $\ddot{n}$ denote the α-integer $\alpha(n)$ and the β-integer $\beta(n)$, respectively.

Henceforth a $\underline{point}$ is a $*$-point, that is an ordered pair of numbers in A and B, in that order. The symbol P_i represents the point (a_i, b_i).

10.2 $*$-SPACE

By $*$-space we mean the system consisting of all points and the following method of computing the $\underline{*\text{-distance}}$, $\overset{*}{d}(P_1, P_2)$, between points P_1 and P_2:

$$\overset{*}{d}(P_1, P_2) = \ddot{\sqrt{[\iota(a_1 \dot{-} a_2)]^{\ddot{2}} \ddot{+} [b_1 \ddot{-} b_2]^{\ddot{2}}}}$$

75

$$= \beta\left(\sqrt{[\alpha^{-1}(a_1) - \alpha^{-1}(a_2)]^2 + [\beta^{-1}(b_1) - \beta^{-1}(b_2)]^2}\right).$$

This *-distance formula gives the same results informally explained in Section 6.11 and was initially designed to facilitate the graphic interpretation presented there. (When $\alpha = \beta = I$, *-distance is obviously the usual Euclidean distance.) Observe that $\overset{*}{d}(P_1,P_2)$ is in B, as are the values of all the operators of the *-calculus. Of course, one could easily redefine *-distance so that it is in A, but that would be less useful for our purposes.

If $a_1 = a_2$, then $\overset{*}{d}(P_1,P_2) = |b_1 \overset{..}{-} b_2|$; and if $b_1 = b_2$, then $\overset{*}{d}(P_1,P_2) = |\iota(a_1 \overset{.}{-} a_2)| = \iota(|a_1 \overset{.}{-} a_2|)$.

Although $\overset{*}{d}$ is not a metric in the usual sense, it is a *-metric in the sense that the following proposition is true.

For any points P_1, P_2, and P_3,

$$\overset{*}{d}(P_1,P_2) \overset{..}{\geq} \overset{..}{0},$$

$$\overset{*}{d}(P_1,P_2) = \overset{..}{0} \text{ if and only if } P_1 = P_2,$$

$$\overset{*}{d}(P_1,P_2) = \overset{*}{d}(P_2,P_1),$$

$$\overset{*}{d}(P_1,P_2) \overset{..}{+} \overset{*}{d}(P_2,P_3) \overset{..}{\geq} \overset{*}{d}(P_1,P_3).$$

The points P_1, P_2, and P_3 are *-collinear provided that at least one of the following holds:

$$\overset{*}{d}(P_2,P_1) \overset{..}{+} \overset{*}{d}(P_1,P_3) = \overset{*}{d}(P_2,P_3),$$

$$\overset{*}{d}(P_1,P_2) \overset{..}{+} \overset{*}{d}(P_2,P_3) = \overset{*}{d}(P_1,P_3),$$

$$\overset{*}{d}(P_1,P_3) \overset{..}{+} \overset{*}{d}(P_3,P_2) = \overset{*}{d}(P_1,P_2).$$

If $a_1 = a_2 = a_3$ or $b_1 = b_2 = b_3$, then P_1, P_2, and P_3 are *-collinear, but those are not the only circumstances that engender *-collinearity, as we shall see shortly.

A *-line (or *-geodesic) is a set L of at least two distinct points such that for any distinct points P_1 and P_2 in

L, a point P_3 is in L if and only if P_1, P_2, and P_3 are *-collinear. A *-line is <u>vertical</u> provided all its points have the same first member.

When $\alpha = \beta = I$, the *-lines are the straight lines of Euclidean analytic geometry in two dimensions. If $\alpha = I$ and $\beta(x) = x^{\overset{=}{2}}$, for example, then each nonvertical *-line is the union of two parabolic parts. (The parabolic family of calculi is defined in Note 6.)

<u>Theorem</u>

> The class of nonvertical *-lines is identical with the class of *-uniform functions.

Two *-lines are <u>*-parallel</u> provided they are identical or have no common point. It turns out that two nonvertical *-lines are *-parallel if and only if they have the same *-slope.

A *-line L_1 is <u>*-perpendicular</u> to a *-line L_2 provided they have a common point P and for any points P_1 on L_1 and P_2 on L_2, distinct from P, $\overset{*}{d}(P_1,P) \overset{*}{<} \overset{*}{d}(P_1,P_2)$. If L_1 and L_2 are *-perpendicular, then for any points P_1 on L_1 and P_2 on L_2,

$$[\overset{*}{d}(P,P_1)]^{\overset{..}{2}} \overset{..}{+} [\overset{*}{d}(P,P_2)]^{\overset{..}{2}} = [\overset{*}{d}(P_1,P_2)]^{\overset{..}{2}}.$$

Furthermore, two nonvertical *-lines are *-perpendicular if and only if the β-product of their *-slopes is $\overset{..}{-}1$, that is, $\beta(-1)$.

The <u>*-arc length</u> of a suitably behaved arc can be defined in an obvious way by partitioning the arc and using the *-distance between the successive points. If f has a *-continuous *-derivative on $[r,s]$, the *-arc length of f on $[r,s]$ turns out to be

$$\overset{*}{\int_r^s} \overset{..}{\sqrt{\overset{..}{1} \overset{..}{+} [\overset{*}{D}f]^{\overset{..}{2}}}}.$$

As expected, the *-distance between two points P_1 and P_2 is equal to the *-arc length of the *-segment connecting P_1 and P_2.

With the *-metric one can define all the *-conics, the study of which may be facilitated by using the *-vectors defined in the next section. (It would be interesting to know whether one can select α and β so that every *-circle is an ellipse.)

Since *-space is, in fact, a model of plane Euclidean geometry, all the theorems of the latter have counterparts in the former. Nevertheless, we believe that it is just as profitable to distinguish these spaces as it is to distinguish the calculi.

Clearly the foregoing developments can be extended to n dimensions.

In Note 7 we show how to convert the set of all points into a system of *-complex numbers.

10.3 *-VECTORS

For any numbers a in A and b in B, the transformation that maps each point (x,y) into (x∔a, y∺b) is a *-vector and is here denoted by $\overset{*}{v}$[a;b]. The sum of $\overset{*}{v}$[a;b] and $\overset{*}{v}$[c;d] is $\overset{*}{v}$[a∔c, b∺d].

Theorem

> For any point (x,y) and *-vector $\overset{*}{v}$[a;b], the
> point (x,y), its image (x∔a, y∺b), and the
> latter's image (x∔a∔a, y∺b∺b), are *-collinear.

We say that the vector $\overset{*}{v}$[a;b] is rectilinear provided that for EVERY point (x,y), the points (x,y), (x∔a, y∺b), and (x∔a∔a, y∺b∺b) are classically collinear; otherwise the vector is curvilinear.

If $\alpha = \beta = I$, then all *-vectors are rectilinear. But, for instance, if $\alpha = I$ and $\beta = \exp$, then most *-vectors are curvilinear; e.g., $\overset{*}{v}$[7,2] is curvilinear since it maps (1,3) into (8,6), which is mapped into (15,12), three *-collinear points that are not classically collinear.

Various types of scalar multiplications and norms can be defined for *-vectors.

10.4 THE *-METHOD OF LEAST SQUARES

The *-method of least squares is the same kind of extension of the classical method of least squares as the *-calculus is of the classical calculus.

Let f be a discrete function with arguments $a_1, \ldots, a_n$ in A and values in B, and let S be a set of functions whose values are in B and whose arguments are in A and include $a_1, \ldots, a_n$.

If there exists a member g of S such that for every member h of S distinct from g,

$$[g(a_1) \mathbin{\ddot{-}} f(a_1)]^2 \mathbin{\ddot{+}} \ldots \mathbin{\ddot{+}} [g(a_n) \mathbin{\ddot{-}} f(a_n)]^2$$

$$\mathbin{\ddot{<}} \quad [h(a_1) \mathbin{\ddot{-}} f(a_1)]^2 \mathbin{\ddot{+}} \ldots \mathbin{\ddot{+}} [h(a_n) \mathbin{\ddot{-}} f(a_n)]^2,$$

then g, which is obviously unique, is said to be the member of S that is best fitted to f by the *-method and is denoted by $\overset{*}{L}(S;f)$.

When $\alpha = \beta = I$, the *-method is identical with the classical method and the notation $L(S;f)$ will be used instead of $\overset{*}{L}(S;f)$. Please note that $L(S;f)$ is not necessarily a linear function.

For each function θ with arguments in A and values in B, let $\bar{\theta}$ be the function consisting of all ordered pairs $(\alpha^{-1}(x), \beta^{-1}(y))$ for which (x,y) is in θ. Let $\bar{S}$ be the set of all functions $\bar{\theta}$ for which θ is in S. Then $\overset{*}{L}(S;f)$ and $L(\bar{S};\bar{f})$ coexist, and if they do exist,

$$\overset{*}{L}(S;f) = \beta\{L(\bar{S};\bar{f})\}.$$

The pattern of that relationship is already quite familiar in view of the results given in Section 6.8.

Thus, if $\overset{*}{L}(S;f)$ exists, it can be found by using the classical method to select the member of $\bar{S}$ that best fits $\bar{f}$ and then applying β to the result.

Any statistical justification for the classical method can probably be converted into a justification for the *-method.

If S is the set of all *-uniform functions, then we have the following results.

1. $\overset{*}{L}(S;f)$ exists (and is unique).

2. $\{[\overset{*}{L}(S;f)](a_1) \mathbin{\ddot{-}} f(a_1)\} \mathbin{\ddot{+}} \ldots$

$$\mathbin{\ddot{+}} \{[\overset{*}{L}(S;f)](a_n) \mathbin{\ddot{-}} f(a_n)\} = \ddot{0}.$$

3. $\overset{*}{\text{L}}$(S;f) contains the *-<u>centroid</u> of f, which we
define to be the point whose members are the
α-average of the a_i and the β-average of the
$f(a_i)$, in that order.

In this situation, the *-slope of $\overset{*}{\text{L}}$(S;f) may be useful for
indicating the "overall *-direction" of the discrete func-
tion f.

Consider now the problem of fitting a function of the
form exp(mx + c) to a positive discrete function f:{(a_1,b_1),
..., (a_n,b_n)}. Since the classical method is extremely diffi-
cult to apply here, scientists use the following "lineari-
zation" technique.

First f is transformed into {$(a_1,\ln(b_1))$,...,$(a_n,\ln(b_n))$},
and the functions exp(mx + c) are transformed into ln[exp(mx+
c)], that is, into the linear functions mx + c. Then the
classical method is used to fit a linear function to the
transformation of f, and finally exp is applied to that linear function. The result, which we have not seen explained
heretofore, is identical with that provided by the *-method
for which α = I and β = exp.

The idea for the *-method arose from the following analogue of a well-known theorem concerning the arithmetic average.

If b is the β-average of n numbers $b_1,\ldots,b_n$ in B,
then

$$(b \mathbin{\ddot{-}} b_1) \mathbin{\ddot{+}} \ldots \mathbin{\ddot{+}} (b \mathbin{\ddot{-}} b_n) = \ddot{0},$$

and for any number x in B distinct from b,

$$(b \mathbin{\ddot{-}} b_1)^{\ddot{2}} \mathbin{\ddot{+}} \ldots \mathbin{\ddot{+}} (b \mathbin{\ddot{-}} b_n)^{\ddot{2}}$$

$$\mathbin{\ddot{<}} (x \mathbin{\ddot{-}} b_1)^{\ddot{2}} \mathbin{\ddot{+}} \ldots \mathbin{\ddot{+}} (x \mathbin{\ddot{-}} b_n)^{\ddot{2}}.$$

Perhaps other statistical concepts can be developed in
the context of *-calculus. For example, since the *-graph
(Section 6.11) of the function $\beta\{\exp[-(\alpha^{-1}(x))^2]\}$ is a normal distribution curve, we may call that function *-normal.
If a function is not classically normal, it may be *-normal
for suitable choices of α and β, in which case the tools of
the *-calculus would be available for analytic purposes.

10.5 TRENDS

Although least squares methods provide interesting ways of determining various "overall directions" of a discrete function, there are other methods that are easier to calculate and may prove to be more useful. For simplicity we begin with functions defined on (closed) intervals, rather than with discrete functions.

By a <u>trend</u> of a function on an interval we mean an average of a derivative of the function on the interval. (It may be helpful to conceive of a trend as a kind of global derivative.)

For example, one could compute the quadratic average of the classical derivative of g on [r,s]:

$$\left\{ \frac{1}{s-r} \int_r^s (Dg)^{\overline{\overline{2}}} \right\}^{\overline{\overline{1/2}}} .$$

Or one could compute the biquadratic average of the anageometric derivative of g on a positive interval.

Trends that do not depend on all the points of the function are trivial. For example, the arithmetic average of Dg on [r,s] is a trivial trend, since it equals

$$[g(s) - g(r)]/(s - r),$$

which depends solely on the points $(r,g(r))$ and $(s,g(s))$. Similarly, the *-average of $\overset{*}{D}g$ on $[r,s]$ is a trivial trend. Nontrivial trends are obtained only by applying an average from one calculus to a derivative from another calculus.

One can compute trends of a discrete function f by averaging the slopes of the uniform functions joining successive points of f, assuming, of course, that the arguments of f are equispaced relative to the arithmetic used for the arguments. For instance, one could compute the quadratic average of the classical slopes of the linear functions connecting the successive points. (The resulting trend is related to the root-mean-square-successive-difference used by some statisticians.)

We would like to know whether there is a trend operator that assigns to each discrete function with equispaced arguments the classical slope of the linear function that can be fitted to f by the classical method of least squares.

10.6 CALCULUS IN BANACH SPACES

Each arithmetic can be converted into a Banach space. For example, α-arithmetic is so converted by defining the norm of each number u in A to be $|\alpha^{-1}(u)|$ and by defining the scalar product of a number r and u to be $\alpha(r) \stackrel{\times}{\cdot} u$.

Shortly after completing our development of the non-Newtonian calculi in August of 1970, we had occasion to consult the first edition of Jean Dieudonné's <u>Foundations</u> <u>of</u> <u>Modern</u> <u>Analysis</u>. There we found a definition of the derivative of a function with arguments in a Banach space and values in a Banach space. If those Banach spaces are taken to be converted α-arithmetic and converted β-arithmetic, respectively, then that derivative is the *-derivative. For the integral, Dieudonné apparently restricted his attention to functions with arguments in R and values in a Banach space. If the Banach space is taken to be converted β-arithmetic, then that integral is the special *-integral for which $\alpha = I$.

However, since we have nowhere seen a discussion of even one specific non-Newtonian calculus, and since we have not found a notion that encompasses the *-average, we are inclined to the view that the non-Newtonian calculi have not been known and recognized heretofore. But only the mathematical community can decide that.

10.7 CONCLUSION

We trust that the basic ideas of the non-Newtonian calculi have been presented in sufficient detail to enable interested persons to develop the theory in various directions. For example, one could study *-differential equations, investigate Taylor series in the context of *-calculus, or construct non-Newtonian calculi of functions of two or more real variables by choosing an arithmetic for each axis. It might even be profitable to seek deeper connections among the corresponding operators of the calculi. One well-known theorem about averages comes to mind:

Let $\alpha = I$. For each positive integer n, let

$$\beta_n(x) = \begin{cases} x^n & \text{if } x \geq 0 \\ x & \text{if } x < 0 \end{cases},$$

and let $*_n$ be (α-arithmetic, β_n-arithmetic). Then for any positive continuous function f on [r,s],

$$\lim_{n \to \infty} \left[\overset{*n}{M} \right]_r^s f = \widetilde{M}_r^s f.$$

If the corresponding propositions for $*_n$-derivatives and integrals were true, then one could say that the geometric calculus is the "limit" of an infinite sequence of calculi. But, alas, those propositions are false. Nevertheless, it is not unreasonable to speculate that there are nontrivial sequences of calculi which do converge in some well-defined sense.

Some years ago we encountered a remark by N.J. Lennes that has become for us a constant reminder of the hazards of premature judgment:

"Christiaan Huygens, one of the world's great mathematicians and physicists, ... never became enthusiastic about the calculus, and urged to the end of his life that all problems solved by means of it could be solved equally well by the older methods."

* * *

N O T E S

Note 1 (to page 12)

Those who prefer percents to ratios may find the following concepts more appealing than the corresponding concepts of the geometric calculus. However, the relative derivative is not a non-Newtonian derivative according to our use of the term and there is no integral that is the 'inverse' of the relative derivative. It should also be mentioned that the relative derivative is not the so-called logarithmic derivative, which also fails to be a non-Newtonian derivative.

The relative change of a positive function f over [r,s] is [f(s) - f(r)]/f(r), which equals 1 less than the geometric change of f over [r,s]. The relative slope of a geometrically-uniform function is its relative change over any interval of classical extent 1. The relative gradient of a positive function f over [r,s] is the relative slope of the geometrically-uniform function containing (r,f(r)) and (s,f(s)), and turns out to be

$$\left\{ \frac{f(s)}{f(r)} \right\}^{\frac{1}{s-r}} - 1,$$

or 1 less than the geometric gradient of f over [r,s].

In securities analysis and other applications the relative gradient is often called the compound growth rate. For example, suppose that one paid \$64 for a share of stock. Three years later the price of the share is \$216. The compound (annual) growth rate of the price over the time interval [0,3] is given by

$$[216/64]^{1/(3-0)} - 1,$$

which equals 0.5 or 50%. The significance of that figure stems from the fact that if an original investment of \$64 increased 50% in each of three successive years, the final value would be \$216.

The relative derivative of a positive function f at an argument a is the following limit, if it exists and is greater than -1:

84

$$\lim_{x \to a} \left[\left(\frac{f(x)}{f(a)} \right)^{\frac{1}{x-a}} - 1 \right].$$

The relative derivative of f at a coexists with, and equals 1 less than, the geometric derivative of f at a.

Note 2 (to page 13)

We shall explain here how a simple algebraic identity led us to construct the geometric calculus. First we re-express the discrete analogue of the Basic Theorem of Classical Calculus (p.6) in a slightly different manner.

Consider any n points $(a_1, v_1), \ldots, (a_n, v_n)$, where $a_1 < a_2 < \ldots < a_n$ and $a_2 - a_1 = a_3 - a_2 = \ldots = a_n - a_{n-1}$. Let k be the common value of the $a_i - a_{i-1}$, so that $k(n-1) = a_n - a_1$. Connect the n points by line segments, of which there are n - 1. Then the arithmetic average of their classical slopes equals the classical slope of the line segment containing the endpoints (a_1, v_1) and (a_n, v_n); that is

$$\frac{\dfrac{v_2 - v_1}{k} + \dfrac{v_3 - v_2}{k} + \ldots + \dfrac{v_n - v_{n-1}}{k}}{n - 1} = \frac{v_n - v_1}{a_n - a_1}.$$

Now assume that the v_i are positive. Then the following identity is obvious:

$$\left\{ \left[\frac{v_2}{v_1} \right]^{\frac{1}{k}} \cdot \left[\frac{v_3}{v_2} \right]^{\frac{1}{k}} \cdots \left[\frac{v_n}{v_{n-1}} \right]^{\frac{1}{k}} \right\}^{\frac{1}{n-1}} = \left[\frac{v_n}{v_1} \right]^{\frac{1}{a_n - a_1}}$$

The left side of that equation is the geometric average of the n-1 numbers $(v_i/v_{i-1})^{1/k}$, which, we imagined, must be slopes of a new kind. We conjectured that the preceding equation must be the discrete analogue of the Basic Theorem of a new system of calculus, the geometric calculus.

Note 3 (to page 19)

Here we explain why, in the definition of anageometric slope, we use positive intervals of geometric extent equal to e, rather than to some other number greater than 1. (Geometric extent is necessarily greater than 1.)

Let w > 1, and assume we had defined the anageometric slope of an anageometrically-uniform function to be its classical change over any positive interval of geometric extent w. Then the anageometric slope of $\ln(cx^m)$, the typical anageometrically-uniform function, would equal $m \cdot \ln(w)$. Accordingly, it is convenient to stipulate that $\ln(w) = 1$, or $w = e$. The choice of $w = e$ is no more arbitrary here than the choice of intervals of classical extent 1 in the definition of classical slope (Section 1.2). In Section 6.10 we make another comment on the choice $w = e$.

Note 4 (to page 49)

Many 'growth' calculi can be constructed by taking $\alpha = I$ and choosing β from the growth curves, which include or are related to the logistic curves, cumulative normal curves, and the sigmoidal curves that occur in the study of population and biological growth.

For example, let $\sigma(x) = (e^x - 1)/(e^x + 1)$. {Note that $\sigma(2x) = \tanh(x)$.} The function σ is of the sigmoid type (S-shaped) and generates sigmoidal arithmetic, whose realm consists of all numbers strictly between -1 and 1. The sigmoidal sum and difference of numbers u and v in the realm turn out to be $(u + v)/(1 + uv)$ and $(u - v)/(1 - uv)$, respectively. The sigmoidal calculus is the $*$-calculus for which $\alpha = I$ and $\beta = \sigma$. By choosing $\alpha = \sigma$ and $\beta = I$ we get the anasigmoidal calculus; for $\alpha = \beta = \sigma$ we obtain the bisigmoidal calculus.

It is entertaining to extend sigmoidal arithmetic by appending -1 and 1 to the realm and defining the sum of numbers u and v to be

$$u \overset{\sigma}{+} v = \begin{cases} (u + v)/(1 + uv) & \text{if } uv \neq -1 \\ 0 & \text{if } uv = -1 \end{cases}.$$

Then $1 \overset{\sigma}{+} (-1) = 0$; if $u \neq -1$, then $u \overset{\sigma}{+} 1 = 1$; and if $u \neq 1$, then $u \overset{\sigma}{+} (-1) = -1$. Thus, -1 and 1 behave as negative and positive infinity in the extended sigmoidal system.

If units are chosen so that the speed of light is 1, then the extended sigmoidal sum of two speeds equals the relativistic composition of the speeds, even if one or both of the speeds is 1.

Note 5 (to pages 53 and 60)

In this Note we show how to generate the infinitely-many power arithmetics, of which the quadratic and harmonic arithmetics are special instances. Let p be an arbitrary nonzero number. The $\overline{\overline{p}}$th-power function is the function that assigns to each number x the number

$$x^{\overline{\overline{p}}} = \begin{cases} x^p & \text{if } x > 0 \\ 0 & \text{if } x = 0 \\ -(-x)^p & \text{if } x < 0 \end{cases}.$$

Note that $x^{\overline{\overline{p}}}$ is positive when x is positive, and negative when x is negative.

For any numbers y and z,

$$(yz)^{\overline{\overline{p}}} = y^{\overline{\overline{p}}} \cdot z^{\overline{\overline{p}}}, \quad \text{and}$$

$$(y^{\overline{\overline{p}}})^{\overline{\overline{1/p}}} = y = (y^{\overline{\overline{1/p}}})^{\overline{\overline{p}}}.$$

If p = 1/2, the $\overline{\overline{p}}$th-power function is the function $x^{\overline{\overline{1/2}}}$, which generates the quadratic arithmetic (Section 7.1); if p = -1, the $\overline{\overline{p}}$th-power function is the function 1⫽x, which generates the harmonic arithmetic (Section 8.1); if p = 2, the $\overline{\overline{p}}$th-power function is the function $x^{\overline{\overline{2}}}$, which generates the parabolic arithmetic (Note 6).

The $\overline{\overline{p}}$th-power function is one-to-one, is on R and onto R, has the $\overline{\overline{1/p}}$th-power function as its inverse, and gener-ates the $\overline{\overline{p}}$th-power arithmetic. In that arithmetic the nat-ural average of n numbers $u_1,\ldots,u_n$ turns out to be

$$\left\{ (u_1^{\overline{\overline{1/p}}} + \ldots + u_n^{\overline{\overline{1/p}}})/n \right\}^{\overline{\overline{p}}},$$

which reduces to the well-known pth-power average when the u_i are positive.

Note 6 (to page 77)

The function $\rho(x) = x^{\overline{\overline{2}}}$ generates the <u>parabolic arithme-tic</u>, which is one of the power arithmetics discussed in Note 5 above. The realm of parabolic arithmetic is R. The <u>para-bolic calculus</u> is the *-calculus for which α = I and β = ρ;

the parabolically-uniform functions are expressible in the form $(mx + c)^2$. By choosing $\alpha = \rho$ and $\beta = I$ we get the <u>anaparabolic</u> <u>calculus</u>; for $\alpha = \beta = \rho$ we obtain the <u>bipara-bolic</u> <u>calculus</u>.

<u>Note 7 (to page 78)</u>

The system of <u>*-complex</u> <u>numbers</u> consists of all *-points and two operations defined thus: the sum of (a_1, b_1) and (a_2, b_2) is $(a_1 \overset{\cdot}{+} a_2,\ b_1 \overset{\cdot\cdot}{+} b_2)$ and their product is

$$\left(\alpha\{\bar{a}_1\bar{a}_2 - \bar{b}_1\bar{b}_2\},\ \beta\{\bar{a}_1\bar{b}_2 + \bar{b}_1\bar{a}_2\} \right),$$

where $\bar{a}_1 = \alpha^{-1}(a_1)$, $\bar{a}_2 = \alpha^{-1}(a_2)$, $\bar{b}_1 = \beta^{-1}(b_1)$, and $\bar{b}_2 = \beta^{-1}(b_2)$.

For this system, which is a field, one can define sub-traction and division that are 'inverses' of addition and multiplication. Furthermore, α-arithmetic (and hence every arithmetic) is embedded in the system.

Since the product of $(\overset{\cdot}{0}, \overset{\cdot\cdot}{1})$ with itself equals $(\overset{\cdot}{-1}, \overset{\cdot\cdot}{0})$, we may define $\overset{*}{i}$ to be $(\overset{\cdot}{0}, \overset{\cdot\cdot}{1})$. Of course, the product of $(\overset{\cdot}{0}, \overset{\cdot\cdot}{-1})$ with itself also equals $(\overset{\cdot}{-1}, \overset{\cdot\cdot}{0})$.

Various types of moduli can be defined for the *-complex numbers.

<u>LIST OF SYMBOLS</u>

Single-dotted notations, such as $\dot{+}$, $\dot{<}$, and $[r,\dot{s}]$, pertain to α-arithmetic; double-dotted notations pertain to β-arithmetic.

A, 32,38; $\bar{A}$, 45

B, 38; $\bar{\bar{B}}$, 45

D, 4; $\tilde{D}$, 11-12; $\underset{\sim}{D}$, 19-20; $\underset{\sim}{\tilde{D}}$, 26; $\overset{*}{D}$, 40; $\overset{\bar{*}}{D}$, 46;
  $\overset{\circ}{D}$, 53; $\underset{\circ}{D}$, 55; $\underset{\circ}{\overset{\circ}{D}}$, 57; $\overset{\square}{D}$, 61; $\underset{\square}{D}$, 63; $\underset{\square}{\overset{\square}{D}}$, 64

G, 16; $\tilde{G}$, 16; $\underset{\sim}{G}$, 24; $\underset{\sim}{\tilde{G}}$, 31; $\overset{*}{G}$, 45; $\overset{\bar{*}}{G}$, 46

I, 1

L(S;f), 79; $\overset{*}{L}$(S;f), 79

M, 5; $\tilde{M}$, 13; $\underset{\sim}{M}$, 20; $\underset{\sim}{\tilde{M}}$, 27; $\overset{*}{M}$, 41; $\overset{\bar{*}}{M}$, 46

P_1, P_2, P_3, 75

R, 1; R_+, 1

α, 33-34, 38

α^{-1} is the inverse of α. (Page 1)

β, 38

β^{-1} is the inverse of β. (Page 1)

ι, 39

$\overset{*}{\nu}$, 78

$[r,s]$, 1; $[r,\dot{s}]$, 35; $\overset{\square}{[r},\overset{\square}{s]}$, 59

e, 1

exp, 1

ln, 1-2

$\dot{+}$, 32-34; $\ddot{+}$, 38; $\oplus$, 52; $\boxplus$, 59

$\dot{-}$, 32-34; $\ddot{-}$, 38; $\ominus$, 52; $\boxminus$, 59

$\dot{\times}$, 32-34; $\ddot{\times}$, 38

$\dot{/}$, 32-34; $\ddot{/}$, 38

$\dot{<}$, 32-34; $\ddot{<}$, 38; $\boxed{<}$, 59

$\dot{0}$, 33, 34; $\ddot{0}$, 39, 75

$\dot{1}$, 33, 34; $\ddot{1}$, 39, 75

$\dot{n}$, 34; $\ddot{n}$, 39, 75

$\dot{|x|}$, 34, 75; $\ddot{|x|}$, 75

$*$, 38; $\bar{*}$, 45

$*$-lim, 38; $\bar{*}$-lim, 46

$\overset{*}{d}(P_1, P_2)$, 49, 75-76

$x^{\bar{\bar{2}}}$, 52

$x^{\overline{\overline{1/2}}}$, 52

$\dfrac{1}{x}^{=}$, $1/\!/x$, 59

$b^{\ddot{2}}$, 75

$\ddot{\sqrt{}}$, 75

$\int$,7; $\tilde{\int}$,14-15; $\underset{\sim}{\int}$,22; $\underset{\tilde{\sim}}{\tilde{\int}}$,29; $\overset{*}{\int}$,43; $\overset{\bar{*}}{\int}$,46; $\overset{\circ}{\int}$,54;

$\underset{\circ}{\int}$,56; $\overset{\circ}{\underset{\circ}{\int}}$,58; $\overset{\square}{\int}$,61; $\underset{\square}{\int}$,63; $\overset{\square}{\underset{\square}{\int}}$,65

<u>I N D E X</u>

Most items are listed according to their qualifying
prefixes (α,β,*) or qualifying adjectives (anageometric, anaharmonic, anaquadratic, bigeometric, biharmonic, biquadratic, classical, geometric, harmonic, quadratic).

BIBLIOGRAPHY

Grossman, M.. *The First Nonlinear System of Differential and Integral Calculus*. Rockport, MA: Mathco, 1979.
This book contains a detailed account of the geometric calculus, which was the first of the non-Newtonian calculi. Also included are discussions of the analogy that led to the discovery of that calculus, and some heuristic guides for its application.

Meginniss, J. R. "Non-Newtonian Calculus Applied to Probability, Utility, and Bayesian Analysis." *Proceedings of the American Statistical Association*: Business and Economics Statistics Section (1980), pp. 405-410.
This paper presents a new theory of probability suitable for the analysis of human behavior and decision making. The theory is based on the idea that subjective probability is governed by the laws of a non-Newtonian calculus and one of its corresponding arithmetics.

Grossman, J., Grossman, M., and Katz, R. *The First Systems of Weighted Differential and Integral Calculus*. Rockport, MA: Archimedes Foundation, 1980.
This monograph reveals how weighted averages, Stieltjes integrals, and derivatives of one function with respect to another can be linked to form systems of calculus, which are called weighted calculi because in each such system a weight function plays a central role.

Grossman, J. *Meta-Calculus: Differential and Integral*. Rockport, MA: Archimedes Foundation, 1981.
This monograph contains a development of systems of calculus, called meta-calculi, that transcend the classical calculus, for example in the following manner. In each meta-calculus the gradient, or average rate of change, of a function f on an interval [r,s] depends on *all* the points (x,f(x)) for which $r \leq x \leq s$, whereas the classical gradient $[f(s) - f(r)]/(s - r)$ depends only on the endpoints (r,f(r)) and (s,f(s)). The meta-calculi arose from the problem of measuring stock-price performance when taking all intermediate prices into account.

Grossman, M. *Bigeometric Calculus: A System with a Scale-Free Derivative*. Rockport, MA: Archimedes Foundation, 1983.
This book contains a detailed treatment of the bigeometric calculus, which has a derivative that is scale-free, i.e., invariant under all changes of scales (or units) in function arguments and values. Also included are heuristic guides for the application of that calculus, and various related matters such as the bigeometric method of least squares.

Grossman, J., Grossman, M., and Katz, R. *Averages: A New Approach*. Rockport, MA: Archimedes Foundation, 1983.
This monograph is primarily concerned with a comprehensive family of averages (unweighted and weighted) of functions that arose naturally in the development of non-Newtonian calculus and weighted non-Newtonian calculus. The monograph also contains discussions of some heuristic guides for the appropriate use of averages and an interesting family of means of two positive numbers.